SPRINGER
LAB MANUAL

Springer

Berlin
Heidelberg
New York
Barcelona
Budapest
Hong Kong
London
Milan
Paris
Tokyo

G. Isenberg

Cytoskeleton Proteins

A Purification Manual

 Springer

Prof. Gerhard Isenberg
Technische Universität München
Lehrstuhl für Biophysik E22
James-Franck-Straße
D-85747 Garching bei München
Germany

Library of Congress Cataloging-in-Publication Data

Isenberg, Gerhard, 1951-
    Cytoskeleton proteins : a purification manual / Gerhard Isenberg.
        p.   cm.
    Includes bibliographical references and index.
    ISBN 978-3-642-48961-7   ISBN 978-3-642-79632-6 (eBook)
    DOI 10.1007/978-3-642-79632-6

    1. Cytoskeletal proteins--Purification--Laboratory manuals.
    I. Title.
    QP552.C96I83   1995
    574.19'245'072--dc20   95-21495
                                                                CIP

The use of general descriptive names, registered names, trademarks, etc. in this publication does not imply, even in the absence of a specific statement, that such names are exempt from the relevant protective laws and regulations and therefore free for general use.
Cover design: Struve & Partner, Heidelberg
Typesetting:Camera ready by author
SPIN 10471091   27/3137-5 4 3 2 1 0 - Printed on acid-free paper

A General

B Purification Protocols

C Suppliers List

Index

I User´s Instructions

When working with several cytoskeleton proteins at the same time, we noticed that it would be convenient to have straight forward purification protocols available at the lab bench. These mainstream guidelines provide advice for rapid experimentation, time-saving preparation of buffers, acquisition of the right equipment and ordering chemicals in advance. All in all, this information along with the correct applications, will help today´s lab worker in his or her field to save one of the most important luxuries: *time*.

When starting the purification of a new protein, usually one has to screen the citation index and go to the library, only to find out that just the issue you need is temporarily unavailable. After finally getting the article of interest, you will soon find the important message: "as published previously" or "as documented elsewhere". Consequently, one has to start over again, leaving the beginner with several volumes of journals with enough but often even contradictory information before being able to set up a purification scheme in bits and pieces.

The purification protocols presented in this manual have been selected according to the following criteria: they are either the most efficient, the most economic or the only available techniques. Only proteins which can be obtained in reasonable quantities in a pure form are listed. Proteins that are characterized by antibodies only and for which a routine procedure has not yet been developed have been omitted, and will be possibly listed in the forthcoming issue of this manual. Therefore:

Don´t forget to send the purification protocol of your new protein!

In most cases, specific equipment, and when indicated, individual chemicals or chromatographic media are recommended due to their efficiency and practicability. The "suppliers list" offers the possibility to order the items of interest fast and reliably.

Citations are limited to avoid a long reference list. Moreover, the given references do not necessarily represent the first and original publication; however, the original purifier of the protein is given, unless an improved purification protocol has been developed after the initially published procedure.

The Gen-Bank index is based on the EMBL accession numbers.

A list of abbreviations commonly used in the purification protocols is included in the user´s instructions. A short characterization of each protein describing some main features and including the molecular weights is provided at the beginning of each purification protocol. More detailed information concerning molecular design, regulation, modification and cell biological function may be obtained from the original literature, review articles and serial articles as published in *Methods of Enzymology*.

A time schedule has not been included because of the great variability among the experimenters. However, temperature is a sensitive parameter and care should be taken when following the instructions.

Munich, March 1995   G. Isenberg

Technical University of Munich
Biophysics Department E22
85747 Garching
Germany

II General Lab Equipment

The following general equipment is required for routine cytoskeleton protein purifications:

- a cold room operating at 4° C
- homogenization equipment, i.e.

> Waring blender (tissues)
> Omni-mixer (tissues)
> Polytron tissue homogenizer (model PT 10-35
> with PTA 20S probe and PCU/11 power
> control unit
> Dounce glass/Teflon or glass/glass homogenizer
> with loose (L) or tight (T) fitting
> Branson sonifier (ultrasonification)
> Parr-bomb (cells)

- centrifuges:

> (low, medium and high speed) with appropriate
> rotors (incl. GSA), at least one, or for more
> convenience, two ultracentrifuges (Beckman
> Instr.), with the following appropriate rotors:
> 45-Ti, 60-Ti, 2 x SW-28, SW-41
> low-speed table top centrifuge (Eppendorf)

- column equipment as indicated; if necessary, couple to
  FPLC, HPLC (Pharmacia, Waters etc.)

- pumps (LKB), fraction collectors (LKB, Gilford), UV-
  absorbance monitor (280 nm) for liquid chromatography

- SDS-PAGE Mini-Gel equipment

- thermoregulated water bath (37°C, 50°C), boiling water
  bath (100°C)

- ultrafiltration apparatus with various polysulfone cutoff
  filters (Millipore, Bedford, MA)

4

- vacuum concentration device (Schleicher and Schuell) or alternatively a vacuum centrifuge

III Preparation of Buffers

In general, buffers are made up according to two methods:

a) mix a dissolved solution of buffer and adjust the pH by titrating with a strong counterion,

b) mix a solution of buffer from 10 x concentrated stock solutions.

The pH is generally adjusted at room temperature (RT). Buffers are made up freshly; DTT or protease inhibitors are added immediately prior to use.

Care should be taken when choosing the right buffer composition, since Tris or glycine with their amino groups may inhibit protein functions. Also, hydroxyl groups of maleate, citrate or imidazole may be extremely reactive and may interfere with protein structure and function. Some of these shortcomings can be circumvented by using zwitter-ionic buffers. However, one should bear in mind that some of these zwitter-ionic compounds are incompatible with certain cells and tissues (like Hepes) or, for instance, may interfere with a monomer-polymer equilibrium (as PIPES with tubulin polymerization). Also, zwitter-ionic buffers with their ethanol-amine groups interfere with the Lowry protein assay. In these cases, the Bradford method for protein determination should be used.

In any case, the pH range over which the buffer acts effectively should be evaluated prior to its use. The buffering ranges for various zwitter-ionic and non-zwitter-ionic buffers are shown in Fig. 1.

6

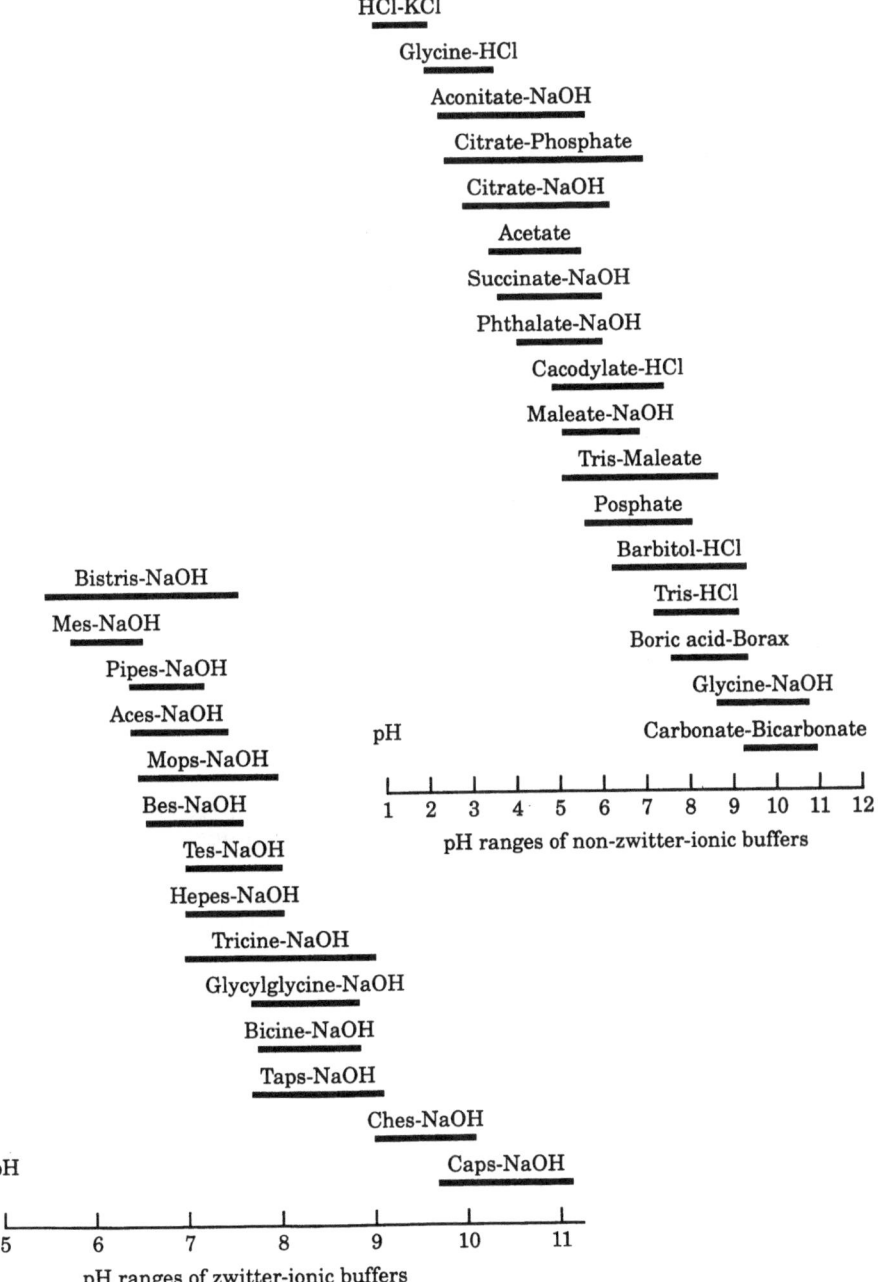

HCl-KCl

Glycine-HCl

Aconitate-NaOH

Citrate-Phosphate

Citrate-NaOH

Acetate

Succinate-NaOH

Phthalate-NaOH

Cacodylate-HCl

Maleate-NaOH

Tris-Maleate

Posphate

Barbitol-HCl

Tris-HCl

Boric acid-Borax

Glycine-NaOH

pH

Carbonate-Bicarbonate

1   2   3   4   5   6   7   8   9   10   11   12

pH ranges of non-zwitter-ionic buffers

Bistris-NaOH

Mes-NaOH

Pipes-NaOH

Aces-NaOH

Mops-NaOH

Bes-NaOH

Tes-NaOH

Hepes-NaOH

Tricine-NaOH

Glycylglycine-NaOH

Bicine-NaOH

Taps-NaOH

Ches-NaOH

Caps-NaOH

pH

5   6   7   8   9   10   11

pH ranges of zwitter-ionic buffers

7

The most common buffers used for cytoskeleton protein preparations

Sodium borate, NaOH buffer (25°C)

Contains 50 ml 0.025 M $Na_2B_4O_7 \cdot 10H_2O$ (9.525 g l^{-1}) and x ml 0.1 M NaOH, diluted to 100 ml

pH	x	β
9.20	0.9	—
9.30	3.6	0.027
9.40	6.2	0.026
9.50	8.8	0.025
9.60	11.1	0.022
9.70	13.1	0.020
9.80	15.0	0.018
9.90	16.7	0.016
10.00	18.3	0.014
10.10	19.5	0.011
10.20	20.5	0.009
10.30	21.3	0.008
10.40	22.1	0.007
10.50	22.7	0.006
10.60	23.3	0.005
10.70	23.8	0.004
10.80	24.25	—

Sodium borate, HCl buffer (25°C)

Contains 50 ml 0.025 M $Na_2B_4O_7 \cdot 10H_2O$ (9.525 g l^{-1}) and x ml 0.1 M HCl, diluted to 100 ml

pH	x	β
8.00	20.5	—
8.10	19.7	0.009
8.20	18.8	0.010
8.30	17.7	0.011
8.40	16.6	0.012
8.50	15.2	0.015
8.60	13.5	0.018
8.70	11.6	0.020
8.80	9.4	0.023
8.90	7.1	0.024
9.00	4.6	0.026
9.10	2.0	—

Glycine, NaOH buffer (25°C)

Contains 25 ml 0.2 M glycine (15.01 g l^{-1}) and x ml 0.2 M NaOH, diluted to 100 ml

pH	x	pH	x
8.6	2.0	9.6	11.2
8.8	3.0	9.8	13.6
9.0	4.4	10.0	16.0
9.2	6.0	10.4	19.3
9.4	8.4	10.6	22.75

Na_2CO_3, $NaHCO_3$ buffer (20°C and 37°C)

Contains x ml 0.1 M $Na_2CO_3 \cdot 10H_2O$ (28.62 g l^{-1}) and (100-x) ml 0.1 M $NaHCO_3$ (8.40 g l^{-1})

pH (25°C)	pH (37°C)	x
9.16	8.77	10
9.40	9.12	20
9.51	9.40	30
9.78	9.50	40
9.90	9.72	50
10.14	9.90	60
10.28	10.08	70
10.53	10.28	80
10.83	10.57	90

NaHCO$_3$, NaOH buffer (25°C)

Contains 50 ml 0.05M NaHCO$_3$ (4.20 g l^{-1}) and x ml 0.1M NaOH, diluted to 100 ml

pH	x	β
9.60	5.0	—
9.70	6.2	0.013
9.80	7.6	0.014
9.90	9.1	0.015
10.00	10.7	0.016
10.10	12.2	0.016
10.20	13.8	0.015
10.30	15.2	0.014
10.40	16.5	0.013
10.50	17.8	0.013
10.60	19.1	0.012
10.70	20.2	0.010
10.80	21.2	0.009
10.90	22.0	0.008
11.00	22.7	—

Ammonia/ammonium chloride buffer (20°C)

Contains x ml 2M NH$_3$ and $(10-x)$ ml 2M NH$_4$Cl, diluted to 100 ml

pH	x	pH	x
8.25	0.50	9.94	7.0
8.61	1.0	10.18	8.0
8.96	2.0	10.51	9.0
9.21	3.0	10.82	9.5
9.58	5.0		

Tris (hydroxymethyl) aminomethane, HCl buffer (25°C)

Contains 50 ml 0.1M Tris (12.114 g l^{-1}) and x ml 0.1M HCl, diluted to 100 ml

pH	x	β
7.00	46.6	0.0085
7.10	45.7	0.010
7.20	44.7	0.012
7.30	43.4	0.013
7.40	42.0	0.015
7.50	40.3	0.017
7.60	38.5	0.018
7.70	36.6	0.020
7.80	34.5	0.023
7.90	32.0	0.027
8.00	29.2	0.029
8.10	26.2	0.031
8.20	22.9	0.031
8.30	19.9	0.029
8.40	17.2	0.026
8.50	14.7	0.024
8.60	12.4	0.022
8.70	10.3	0.020
8.80	8.5	0.016
8.90	7.0	0.014
9.00	5.7	0.011

Citric acid, sodium citrate buffer (23°C)

Contains x ml 0.1M citric acid (21.01 g C$_6$H$_8$O$_7 \cdot$ H$_2$O l^{-1}) and $(50-x)$ ml 0.1M Na$_3$ citrate (29.41g C$_6$H$_5$O$_7$Na$_3$ $\cdot$2H$_2$O l^{-1}), diluted to 100 ml

pH	x	pH	x
3.0	46.5	4.8	23.0
3.2	43.7	5.0	20.5
3.4	40.0	5.2	18.0
3.6	37.0	5.4	16.0
3.8	35.0	5.6	13.7
4.0	33.0	5.8	11.8
4.2	31.5	6.0	9.5
4.4	28.0	6.2	7.2
4.6	25.5		

Maleic acid, Tris, NaOH buffer (23°C)

Contains 25 ml 0.2M Tris and 0.2M maleic acid, and x ml 0.2M NaOH, diluted to 100 ml

pH	x	pH	x
5.2	3.5	7.0	24.0
5.4	5.4	7.2	25.5
5.6	7.75	7.4	27.0
5.8	10.25	7.6	29.0
6.0	13.0	7.8	31.65
6.2	15.75	8.0	34.5
6.4	18.5	8.2	37.5
6.6	21.25	8.4	40.5
6.8	22.9	8.6	43.3

Triethanolamine hydrochloride, NaOH buffer (20°C)

Contains 50 ml 0.1M triethanolamine hydrochloride (18.57 g l^{-1}) and x ml 0.1M NaOH, diluted to 100 ml

pH	x	pH	x
6.8	3.3	7.9	23.8
6.9	4.1	8.0	26.5
7.0	5.1	8.1	29.4
7.1	6.2	8.2	32.1
7.2	7.6	8.3	34.7
7.3	9.2	8.4	37.0
7.4	11.1	8.5	39.1
7.5	13.2	8.6	40.9
7.6	15.5	8.7	42.5
7.7	18.1	8.8	43.9
7.8	20.8		

Citric acid, NaOH buffer (20°C and 50°C)

Contains x ml 2M NaOH, 10 ml 2M citric acid, diluted to 100 ml

x	pH (20°C)	pH (50°C)	x	pH (20°C)	pH (50°C)
1.0	2.15	2.05	16.0	4.37	4.27
2.0	2.39	2.29	17.0	4.50	4.37
3.0	2.58	2.49	18.0	4.62	4.47
4.0	2.75	2.66	19.0	4.74	4.61
5.0	2.89	2.82	20.0	4.87	4.75
6.0	3.04	2.99	21.0	4.98	4.80
7.0	3.18	3.04	22.0	5.11	4.94
8.0	3.32	3.20	23.0	5.21	5.09
9.0	3.46	3.35	24.0	5.34	5.22
10.0	3.59	3.49	25.0	5.49	5.43
11.0	3.75	3.62	26.0	5.63	5.54
12.0	3.90	3.77	27.0	5.80	5.68
13.0	4.03	3.89	28.0	6.02	5.85
14.0	4.14	4.01	29.0	6.33	6.10
15.0	4.27	4.15	29.5	6.51	6.24

Phosphate buffer (25°C)

Contains x ml 0.2 M Na_2HPO_4 † and $(50\text{-}x)$ ml 0.2 M NaH_2PO_4 ‡, diluted to 100 ml

pH	x	pH	x
5.8	4.0	7.0	30.5
6.0	6.15	7.2	36.0
6.2	9.25	7.4	40.5
6.4	13.25	7.6	43.5
6.6	18.75	7.8	45.75
6.8	24.5	8.0	47.35

KH_2PO_4, NaOH buffer (25°C)

Contains 50 ml 0.1 M KH_2PO_4 (13.60 gl⁻¹) and x ml 0.1 M HCl, diluted to 100 ml

pH	x	β
5.80	3.6	—
5.90	4.6	0.010
6.00	5.6	0.011
6.10	6.8	0.012
6.20	8.1	0.015
6.30	9.7	0.017
6.40	11.6	0.021
6.50	13.9	0.024
6.60	16.4	0.027
6.70	19.3	0.030
6.80	22.4	0.033
6.90	25.9	0.033
7.00	29.1	0.031
7.10	32.1	0.028
7.20	34.7	0.025
7.30	37.0	0.022
7.40	39.1	0.020
7.50	40.9	0.016
7.60	42.4	0.013
7.70	43.5	0.011
7.80	44.5	0.009
7.90	45.3	0.008
8.00	46.1	—

Na_2HPO_4, NaOH buffer (25°C)

Contains 50 ml 0.05 M Na_2HPO_4 (7.10 g l⁻¹) and x ml 0.1 M NaOH, diluted to 100 ml

pH	x	β
10.90	3.3	—
11.00	4.1	0.009
11.10	5.1	0.011
11.20	6.3	0.012
11.30	7.6	0.014
11.40	9.1	0.017
11.50	11.1	0.022
11.60	13.5	0.026
11.70	16.2	0.030
11.80	19.4	0.034
11.90	23.0	0.037
12.00	26.9	—

Imidazole, HCl buffer (25°C)

Contains 25 ml 0.2 M imidazole (13.62 g l⁻¹) and x ml 0.2 M HCl, diluted to 100 ml

pH	x	pH	x
6.2	21.45	7.2	9.3
6.4	19.9	7.4	6.8
6.6	17.75	7.6	4.65
6.8	15.2	7.8	3.0
7.0	12.15		

**Sodium pyrophosphate,
HCl buffer (25°C)**

Contains x ml 0.1M HCl and 50 ml
0.1M $Na_4P_2O_7 \cdot 10H_2O$ (44.61 g l^{-1})

pH	x	pH	x
6.90	50.0	8.30	21.4
7.63	40.9	8.67	12.5
7.95	33.3	8.95	8.9
8.10	26.9		

**N-Ethylmorpholine,
HCl buffer (20°C)**

Contains 50 ml 0.2M N-ethylmorpholine (23.03 g l^{-1}) and x ml 0.2M HCl,
diluted to 100 ml

pH	x	pH	x
6.8	46.2	7.8	26.9
7.0	44.2	8.0	20.9
7.2	41.4	8.2	15.4
7.4	37.5	8.4	10.8
7.6	32.6	8.6	7.3

Balanced salt solutions

physiological saline:
    0.9% w/v NaCl

Tris-buffered saline (Tricine-buffered saline):
    0.9% w/v NaCl, 10-50 mM Tris-HCl, pH 7.2

phosphate-buffered saline (PBS):
    NaCl (7.2 g/l), Na_2HPO_4 (1.48 g/l), KH_2PO_4 (0.43 g/l)

Balanced salt solutions

Compound (mg/l)	Dulbecco's	Hank's	Ringer's mammalian	Tyrode's
$CaCl_2$	100	200	250	200
$MgCl_2 \cdot 6H_2O$	100	-	-	100
$MgSO_4 \cdot 7H_2O$	-	200	-	-
KCl	200	400	420	200
KH_2PO_4	200	100	-	-
$NaHCO_3$	-	1273	-	1000
NaCl	8000	8000	6500	8000
Na_2HPO_4	1150	-	-	-
$NaH_2PO_4 \cdot H_2O$	-	-	-	50
glucose	-	2000	1000	-

The following denaturation and precipitation agents are needed to purify cytoskeleton proteins

Formula	Name	MW	Conc	solubility g/100 ml, 20° C
solubilization:				
$C_{12}H_{26}O_4S$ *Na	SDS	273.3	0.1-2%	25
CH_4N_2O	urea	60.6	4-10 M	48-60
CH_6N_3Cl	guanidinium chloride	95.53	4-6 M	76
precipitation:				
$C_2H_4O_2$	acetic acid	60.05	5-10 vol.%	freely miscible
C_3H_6O	acetone	58.08	2-6 vols	"
C_2H_4O	methanol	32.04	1-2 vols	"
C_2H_6O	ethanol	46.1	1-2 vols	"
$CHCl_3$/ $C_5H_{12}O$	chloroform/ isoamylalcohol (24:1)		equal vol.	immiscible with H_2O

To prevent the oxidation if thiol groups, the following reducing agents are used in all preparations

Formula	Name	MW	Conc
C_3H_7NS	cysteine	121.16	-
$C_4H_{10}O_2S_2$	DTT	154.25	1-5 mM
$C_4H_{10}O_2S_2$	DTE	154.25	1-5 mM
C_2H_6OS	2-mercapto-ethanol	78.1	0.1-2%

For the denaturation of proteins or the solubilization of membrane-anchored cytoskeleton proteins, some of the most common detergents are used

Name	MW	HLB	CMC (µM)	Agg. No.
anionic				
sarkosyl	295.4	-	-	-
SDS	288.4	~40	8200	620
zwitter-ionic				
CHAPS	614.9	-	$6-10 \times 10^3$	10
non-ionic				
Brij 58	1122	15.7	77	40
Octyl-β-D-gluco-pyranoside	292.4	-	215×10^3	84
Triton X-100	625	13.5	250	100-155
Triton X-114	537	12.4	-	-
Nonidet P-40	625	13.1	250	100-155
Tween-20	1228	10.7	59	-
Tween 80	1310	15.0	12	58

HLB (hydrophilic-lipophilic balance number) is a direct measurable number of the hydrophilic character of the detergent: HLB is large with increasing hydrophilicity.

CMC (critical micellization concentration).

Aggregation No. (Agg. No.) describes micellar composition over time and is calculated as $\text{Agg. No.} = \dfrac{\text{micellar MW}}{\text{monomeric MW}}$.

Detergents may be purchased from Calbiochem, San Diego, CA.

IV Ammonium Sulfate Precipitation

Ammonium sulfate precipitation is an appropriate method to fractionate proteins prior to purification by chromatographic procedures. It is also used for concentrating protein samples. Ultra-pure ammonium sulfate (e.g. supplied by Schwarz-Mann) is recommended. AS may be added as solid compound or by dilution from a saturated stock solution.

Saturated ammonium sulfate solutions at various temperatures

	Temperature (°C)				
	0	10	20	25	30
Percentage w/w	41.42	42.22	43.09	43.47	43.85
Gram required to saturate 1 liter	706.80	730.50	755.80	766.80	777.50
Gram in a liter of saturated solution	514.70	525.10	536.10	541.20	545.90
Molarity of saturated solution	3.90	3.97	4.06	4.10	4.13

Chart for ammonium sulfate precipitation of proteins

Initial conc of ammonium sulfate, % saturation at 0°C	Final conc of ammonium sulfate, % saturation at 0°C																
	20	25	30	35	40	45	50	55	60	65	70	75	80	85	90	95	100
	g solid ammonium sulfate to add to 100 ml of solution																
0	10.7	13.6	16.6	19.7	22.9	26.2	29.5	33.1	36.6	40.4	44.2	48.3	52.3	56.7	61.1	65.9	70.7
5	8.0	10.9	13.9	16.8	20.0	23.2	26.6	30.0	33.6	37.3	41.1	45.0	49.1	53.3	57.8	62.4	67.1
10	5.4	8.2	11.1	14.1	17.1	20.3	23.6	27.0	30.5	34.2	37.9	41.8	45.8	50.0	54.4	58.9	63.6
15	2.6	5.5	8.3	11.3	14.3	17.4	20.7	24.0	27.5	31.0	34.8	38.6	42.6	46.6	51.0	55.5	60.0
20	0	2.7	5.6	8.4	11.5	14.5	17.7	21.0	24.4	28.0	31.6	35.4	39.2	43.3	47.5	51.9	56.5
25		0	2.7	5.7	8.5	11.7	14.8	18.2	21.4	24.8	28.4	32.1	36.0	40.1	44.2	48.5	52.9
30			0	2.8	5.7	8.7	11.9	15.0	18.4	21.7	25.3	28.9	32.8	36.7	40.8	45.1	49.5
35				0	2.8	5.8	8.8	12.0	15.3	18.7	22.1	25.8	29.5	33.4	37.4	41.6	45.9
40					0	2.9	5.9	9.0	12.2	15.5	19.0	22.5	26.2	30.0	34.0	38.1	42.4
45						0	2.9	6.0	9.1	12.5	15.8	19.3	22.9	26.7	30.6	34.7	38.8
50							0	3.0	6.1	9.3	12.7	16.1	19.7	23.3	27.2	31.2	35.3
55								0	3.0	6.2	9.4	12.9	16.3	20.0	23.8	27.7	31.7
60									0	3.0	6.3	9.6	13.1	16.6	20.4	24.2	28.3
65										0	3.1	6.4	9.8	13.4	17.0	20.8	24.7
70											0	3.1	6.6	10.0	13.6	17.3	21.2
75												0	3.2	6.7	10.2	13.9	17.6
80													0	3.2	6.8	10.4	14.1
85														0	3.3	6.9	10.6
90															0	3.4	7.1
95																0	3.5
100																	0

V Chromatographic Fractionation

Gel-filtration, Desalting and Buffer Exchange

Separation depends on molecular size; large molecules will not enter the bed and elute first, whereas smaller molecules will elute according to their differing retention times. The original medium for this purification step is Sephadex or Sepharose for larger molecules. Sephacryl is used because of its superior flow rates. Media may be purchased from Pharmacia.

Ion Exchange Chromatography

The most widely used technique for protein separation on the basis of charges. Low and high pressure liquid chromatography guarantees effective separation with two options.

Hydrophobic Interaction Chromatography

The conditions of low pH and the use of organic solvents favour denaturation. Solute retention occurs due to interactions with stationary hydrocarbon surfaces. A technique of choice at low (Bio Rad, Micro-Prep® series) and high pressure (reversed-phase HPLC, Pharmacia, Waters).

Affinity Chromatography

Molecules are purified by reversible adsorption to immobilized ligands. The choice of medium depends on the type of molecule being separated. For routine use CNBr-activated Sepharose 4B (Pharmacia) or the Affi-Prep 10 series from Bio Rad are recommended.

Hydroxyl-apatide

The unique properties of the crystalline calcium phosphate, low cost and easy handling allow high resolution during initial steps in the purification of proteins. The hydrated bio-gel HT hydroxylapatide gel offered from Bio Rad is the most commonly used resin applicable with high protein precision.

Deionization, Metal Ion and Detergent Removal

Chelex® 100 is a chelating resin that binds polyvalent ions with high selectivity. Bio-Beads SM-2 adsorbents remove hydrophobic compounds as non-ionic detergents. AG® 1-X4 resin is used to remove SDS.

Ion-exchange cellulose media: physical and chemical properties of Whatman cellulose media

Physical form	Functional group	Normal pH range	Small ion capacity (meq dg⁻¹)	Protein capacity (mg dg⁻¹)	Bed volume (mg ml⁻¹)	Amount of exchanger required per litre bed volume (kg)	Packing density (dg ml⁻¹)
Anion exchange media							
Pre-swollen microgranular							
DE-51	Diethylaminoethyl	2-9	0.20-0.25	175	30	1.20	0.17
DE-52	Diethylaminoethyl	2-9.5	0.88-1.08	700	130	0.90	0.19
DE-53	Diethylaminoethyl	2-12	1.8-2.2	750	150	1.05	0.20
CA-52	Quarternary ammonium	2-12	1.0-1.2	750	150	1.20	0.20
Dry microgranular							
DE-32	Diethylaminoethyl	2-9.5	0.88-1.08	700	140	0.24	0.20
Dry fibrous							
DE-23	Diethylaminoethyl	2-9.5	0.88-1.08	425	60	0.19	0.15
Cation exchange media							
Pre-swollen microgranular							
CM-52	Carboxymethyl	3-10	0.90-1.15	1180	210	1.05	0.18
SE-52	Sulphoxyethyl	2-12	0.9-1.1	1300	195	1.05	0.15
SE-53	Sulphoxyethyl	2-12	2.1-2.6	1300	210	1.15	0.16
Dry microgranular							
CM-32	Carboxymethyl	3-10	0.90-1.15	1180	200	0.21	0.17
Dry fibrous							
CM-23	Carboxymethyl	3-10	0.55-0.70	675	85	0.16	0.13
P-11	Orthophosphate	3-10	3.2-5.30	—	—	0.22	0.17
Hydrogen bonding medium							
Pre-swollen microgranular							
HB-1	Hydroxyl	2-12	0	375	—	0.70	0.15
Cell debris remover							
Pre-swollen microgranular							
CDR	Diethylaminoethyl	2-9.5	0.25-0.35	—	—	1.15	0.19

VI Centrifugation

1 High Performance Centrifugation

For all kinds of protein purification work centrifugation is an important tool. Crude material can be separated into different fractions with the help of high performance centrifuges. The *Beckman High Performance Avanti J Series* is ideal because high g-forces (> 75,000 g) are combined with extremely high acceleration and deceleration rates. A brushless drive with extremely high torque guarantees high efficiency and high throughput. A modern ergonomic design, easy operation and CFC-free refrigerant are other important features.

Beckman high performance centrifuges can be used with a variety of fixed angle rotors, special "Lite" rotors to reduce weight, swinging bucket and vertical rotors. Special rotors for continuous flow, zonal centrifugation and elutriation for cell separation are also available.

2 Ultracentrifugation

Beckman OPTIMA ultracentrifuges are very well suited for all kinds of protein work. Their high speeds - up to 90,000 rpm or almost 700,000 g - combined with very exact temperature control, are ideal for protein separations. OPTIMA ultracentrifuges are environmentally safe because no CFCs are used to cool the rotor or drive. Thermoelectric cooling is used for rotor temperature control, and air cools the drive system. OPTIMA ultracentrifuges are easy to use, run at a very low noise level and their heat output is extremely low (1 kW) compared to other centrifuges. OPTIMA XL ultracentrifuges are equipped with a very useful software package to calculate run times, gradient solution, run simulations and other helpful features like rotor library, rotor dimensions, available tubes etc.

The Beckman rotor program (more than 40 different kinds) allows the selection of the right one for your special application. There are fixed angle rotors available to concentrate dilute protein solution. With their high speeds of 90,000 rpm / 694,000 g, run times are very short. Swinging bucket rotors for all kinds of zonal rate separations are available for high speeds or large volumes. Vertical and near-vertical rotors are mainly used for DNA and lipoprotein separations, in some cases, for protein purifications as well.

When only very small samples are used, a Beckman OPTIMA TL or TLX tabletop ultracentrifuge should be selected. All separations performed in floor model ultracentrifuges of the OPTIMA series can be done in these TL / TLX centrifuges on a smaller scale. The volume range is 230 µl per tube up to 5 ml maximum. Fixed angle, swinging bucket, vertical and near-vertical rotors are available for these tabletop OPTIMAs.

VII Abbreviations

ABP	actin binding protein
ADP	adenosine diphosphate
AMP	adenosine monophosphate
AMP-PNP	5´-adenylylimidophosphate
ATP	adenosine triphosphate
ATP-ase	adenosine triphosphatase
BAME	N-α-benzoyl-L-arginine methylester
Bicine	N,N-bis(2-hydroxyethyl)glycine
CHAPS	3-[(3-cholamidopropyl)-dimethylammino]-1-propanesulfonate
CMC	critical micellization concentration
DEAE	diethylaminoethyl
DFP	di-isopropyl fluorophosphate
DMEM	Dulbecco´s modified Eagles Medium
DMF	dimethylformamide
DMSO	dimethylsulfoxide
DNA	deoxyribonucleic acid
DNA-se	deoxyribonuclease
DTE	dithioerythrol
DTT	dithiothreitol
EDTA	ethylenediaminetetraacetic acid
EGTA	ethyleneglycon-bis(β-aminoethylether)N,N,N´,N´--tetraacetic acid
FPLC	fast purification liquid chromatography
GTP	guanosine triphosphate
HEPES	4-(2-hydroxyethyl)piperazine-1-ethanesulfonic acid
HLB	hydrophilic-lipophilic balance number
HPLC	high-pressure liquid chromatography
kDa	kilodalton
mAb	monoclonal antibody
MAP	microtubule-associated protein
MES	2-morpholinoethanesulfonic acid monohydrate
MOPS	3-morpholinopropanesulfonic acid
MTs	microtubules
NAG	N-acetyl-D-glucosamine
P	phosphate
PAGE	polyacrylamide gel electrophoresis
PEG	polyethylene glycol
PBS	phosphate-buffered saline
PIPES	piperazine-1,4-bis(2-ethanesulfonic acid)
PLAA	poly-(L-aspartic acid)
PLN	podophyllotoxin
PMSF	phenylmethylsulfonylfluoride
RNA	ribonucleic acid
RT	room temperature
SDS	sodium dodecyl sulfate
SLS	sodium lauryl sulfate
TAME	p-tosyl-L-arginine methyl ester
TCA	trichloroacetic acid

TES	N-[tris(hydroxymethyl)methyl]-2-aminomethanesulfonic acid
TFA	trifluoroacetic acid
TPCK	N-tosyl-L-phenylalanine chloromethyl-ketone
Tris	Tris(hydroxymethyl)aminomethane
u.p.	ultra pure

Atomic numbers and weights of the elements

Element	Symbol	Atomic number	Atomic weight	Element	Symbol	Atomic number	Atomic weight
Actinium	Ac	89	227.03	Mendelevium	Md	101	255.09
Aluminium	Al	13	26.98	Mercury	Hg	80	200.59
Americium	Am	95	243.06	Molybdenum	Mo	42	96.94
Antimony	Sb	51	121.75	Neodymium	Nd	60	144.24
Argon	Ar	18	39.95	Neon	Ne	10	20.18
Arsenic	As	33	74.92	Neptunium	Np	93	237.05
Astatine	At	85	210.99	Nickel	Ni	28	58.71
Barium	Ba	56	137.34	Niobium	Nb	41	92.91
Berkelium	Bk	97	247.07	Nitrogen	N	7	14.01
Beryllium	Be	4	9.01	Nobelium	No	102	255.00
Bismuth	Bi	83	208.98	Osmium	Os	76	190.20
Boron	B	5	10.81	Oxygen	O	8	16.00
Bromine	Br	35	79.90	Palladium	Pd	46	106.40
Cadmium	Cd	48	112.40	Phosphorus	P	15	30.97
Calcium	Ca	20	40.08	Platinum	Pt	78	195.09
Californium	Cf	98	249.07	Plutonium	Pu	94	242.06
Carbon	C	6	12.01	Polonium	Po	84	208.98
Cerium	Ce	58	140.12	Potassium	K	19	39.10
Cesium	Cs	55	132.91	Praseodymium	Pr	59	140.91
Chlorine	Cl	17	35.45	Promethium	Pm	61	145.00
Chromium	Cr	24	52.00	Protactinium	Pa	91	231.04
Cobalt	Co	27	58.93	Radium	Ra	88	226.03
Copper	Cu	29	63.55	Radon	Rn	86	222.02
Curium	Cm	96	245.07	Rhenium	Re	75	186.20
Dysprosium	Dy	66	162.50	Rhodium	Rh	45	102.91
Einsteinium	Es	99	254.09	Rubidium	Rb	37	85.47
Erbium	Er	68	167.26	Ruthenium	Ru	44	101.07
Europium	Eu	63	151.96	Samarium	Sm	62	150.40
Fermium	Fm	100	252.08	Scandium	Sc	21	44.96
Fluorine	F	9	18.99	Selenium	Se	34	78.96
Francium	Fr	87	223.02	Silicon	Si	14	28.09
Gadolinium	Gd	64	157.25	Silver	Ag	47	107.87
Gallium	Ga	31	69.72	Sodium	Na	11	22.99
Germanium	Ge	32	72.59	Strontium	Sr	38	87.62
Gold	Au	79	196.97	Sulfur	S	16	32.06
Hafnium	Hf	72	178.49	Tantalum	Ta	73	180.95
Helium	He	2	4.00	Technetium	Tc	43	98.91
Holmium	Ho	67	164.93	Tellurium	Te	52	127.60
Hydrogen	H	1	1.01	Terbium	Tb	65	158.93
Indium	In	49	114.82	Thallium	Tl	81	204.37
Iodine	I	53	126.90	Thorium	Th	90	232.04
Iridium	Ir	77	192.22	Thulium	Tm	69	168.93
Iron	Fe	26	55.85	Tin	Sn	50	118.69
Khurchatovium	Kh	104	260.00	Titanium	Ti	22	47.90
Krypton	Kr	36	83.80	Tungsten	W	74	183.85
Lanthanum	La	57	138.91	Uranium	U	92	238.03
Lawrencium	Lr	103	256.00	Vanadium	V	23	50.94
Lead	Pb	82	207.20	Xenon	Xe	54	131.30
Lithium	Li	3	6.94	Ytterbium	Yb	70	173.04
Lutetium	Lu	71	174.97	Yttrium	Y	39	88.91
Magnesium	Mg	12	24.31	Zinc	Zn	30	65.37
Manganese	Mn	25	54,94	Zirconium	Zr	40	91.22

I Actin and Actin-Associated Proteins

240 kDa ACTIN BINDING PROTEIN (ABP-240)

A 240 kDa filamin equivalent protein from Dictyostelium discoideum, which crosslinks and bundles actin filaments at the cell cortex.

Source: Dictyostelium discoideum strain AX-3

Equipment:
- swinging bucket rotor 6.0 liter
- polystyrene motor driven (Teflon-glass) homogenizer (10 ml)
- overhead stirrer
- Parr bomb
- ultracentrifuge
- DEAE-cellulose column (2.5 x 30 cm)
- hydroxylapatide column (1 x 26 cm)
- sephacryl S-300 column (1.6 x 70 cm)
- electrophoretic gels
- vacuum concentrator

Chemicals: HL5-medium, PIPES, EGTA, DTT, $CaCl_2$, ATP, NaN_3, 2-mercaptoethanol, NaCl, NaH_2PO_4

protease inhibitors: trasylol, pepstatin, leupeptin, chymostatin

Have ready: **HL5 medium**

buffer A: 5 mM PIPES, pH 7.0

buffer B: 5 mM EGTA, 1 mM DTT, 5 mM PIPES, pH 7.0

buffer C: 0.1 mM $CaCl_2$, 0.5 mM ATP, 0.75 mM 2-mercaptoethanol, 0.02% (w/v) NaN_3, 10 mM PIPES, pH 7.5

DEAE-column buffer: buffer C+0.1 M NaCl, pH 7.5

buffer D: 0.1 mM $CaCl_2$, 0.5 mM ATP, 0.75 mM 2-mercaptoethanol, 0.02% (w/v) NaN_3, 3 mM PIPES, pH 7.5

buffer G: 500 mM NaCl, 2 mM EGTA, 0.02% (w/v) NaN_3, 5 mM PIPES, pH 6.8

buffer H: 500 mM NaCl, 2 mM EGTA, 0.02% (w/v) NaN_3, 0.1 mM DTT, 5 mM PIPES, pH 6.8

protease inhibitors: 0.03 ml trasylol /ml, 10 mg/ml of pepstatin, leupeptin, chymostatin

Reference: Hock, R. and Condeelis, J. (1987) J. Biol. Chem. 262, 394-400

Dictyostelium discoideum strain AX-3

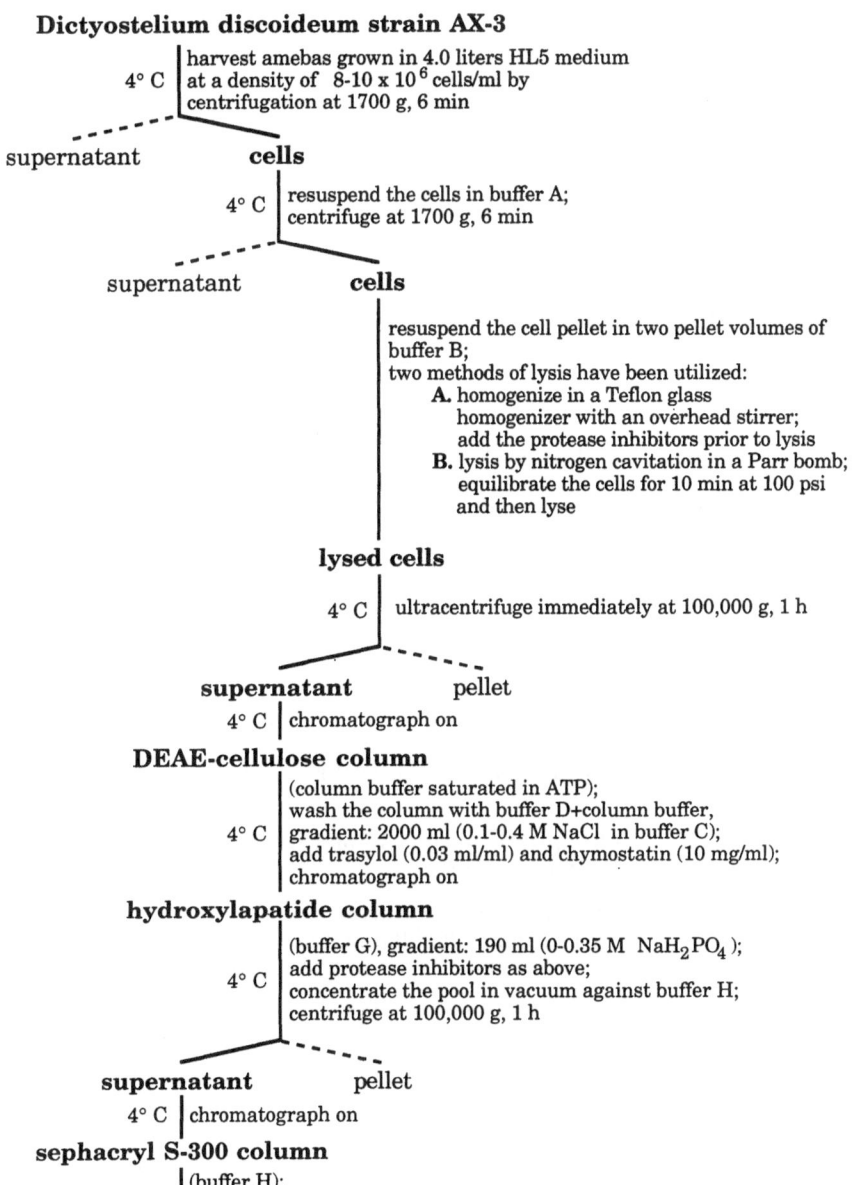

4° C | harvest amebas grown in 4.0 liters HL5 medium
at a density of 8-10 x 10^6 cells/ml by
centrifugation at 1700 g, 6 min

supernatant   **cells**

4° C | resuspend the cells in buffer A;
centrifuge at 1700 g, 6 min

supernatant   **cells**

resuspend the cell pellet in two pellet volumes of
buffer B;
two methods of lysis have been utilized:
   A. homogenize in a Teflon glass
   homogenizer with an overhead stirrer;
   add the protease inhibitors prior to lysis
   B. lysis by nitrogen cavitation in a Parr bomb;
   equilibrate the cells for 10 min at 100 psi
   and then lyse

lysed cells

4° C | ultracentrifuge immediately at 100,000 g, 1 h

supernatant   pellet

4° C | chromatograph on

DEAE-cellulose column

(column buffer saturated in ATP);
wash the column with buffer D+column buffer,
4° C | gradient: 2000 ml (0.1-0.4 M NaCl in buffer C);
add trasylol (0.03 ml/ml) and chymostatin (10 mg/ml);
chromatograph on

hydroxylapatide column

(buffer G), gradient: 190 ml (0-0.35 M NaH$_2$PO$_4$);
4° C | add protease inhibitors as above;
concentrate the pool in vacuum against buffer H;
centrifuge at 100,000 g, 1 h

supernatant   pellet

4° C | chromatograph on

sephacryl S-300 column

(buffer H);
ABP-240 elutes immediately after the void volume

ABP 240

27

30 kDa ACTIN CROSS-LINKING PROTEIN

A monomeric, Ca^{++} inhibited 30 kDa actin bundling and crosslinking protein from Dictyostelium and higher eukaryotes with a molecular contour length of 5.1 nm.

Source: Dictyostelium discoideum axenic strain

Equipment:
- centrifuge
- Parr cell disruption bomb (700 ml)
- glass wool
- dounce homogenizer (50 ml)
- DEAE-cellulose column (1.8 x 10 cm)
- hydroxylapatite column (1.2 x 8 cm)
- hydroxylapatite column (0.5 x 2.5 cm)
- electrophoretic gels

Chemicals: KH_2PO_4, NaH_2PO_4, PIPES, EGTA, EDTA, DTT, Trizma Base, ATP, $CaCl_2$, NaN_3, 2-mercaptoethanol, KCl, $MgCl_2$, NaCl
protease inhibitors: leupeptin, aprotinin, pepstatin, PMSF
column materials: DEAE-cellulose, hydroxylapatide

Have ready: **phosphate buffer:** 15 mM KH_2PO_4, 2 mM NaH_2PO_4, pH 6.2

homogenization buffer: 5 mM PIPES, 5 mM EGTA, 1mM DTT, 0.04 ml aprotinin /ml, 10 mg/ml leupeptin, 1 mg/ml pepstatin, pH 6.9

DEAE-cellulose buffer: 10 mM Tris-HCl, 0.5 mM ATP, 0.5 mM DTT, 0.2 mM $CaCl_2$, pH 7.5

hydroxylapatite buffer: 10 mM PIPES, 0.1 mM EDTA, 0.02% NaN_3, 1.25 mM 2-mercaptoethanol, pH 6.5

storage buffer: 2 mM PIPES, 50 mM KCl, 0.2 mM DTT, 0.02% NaN_3, pH 7.0

Reference: Fechheimer, M. and Furukawa, R. (1991) Methods Enzymol. 196, 84-91
Gen-Bank: M58022

axenic strain of D-discoideum (Ax-3)

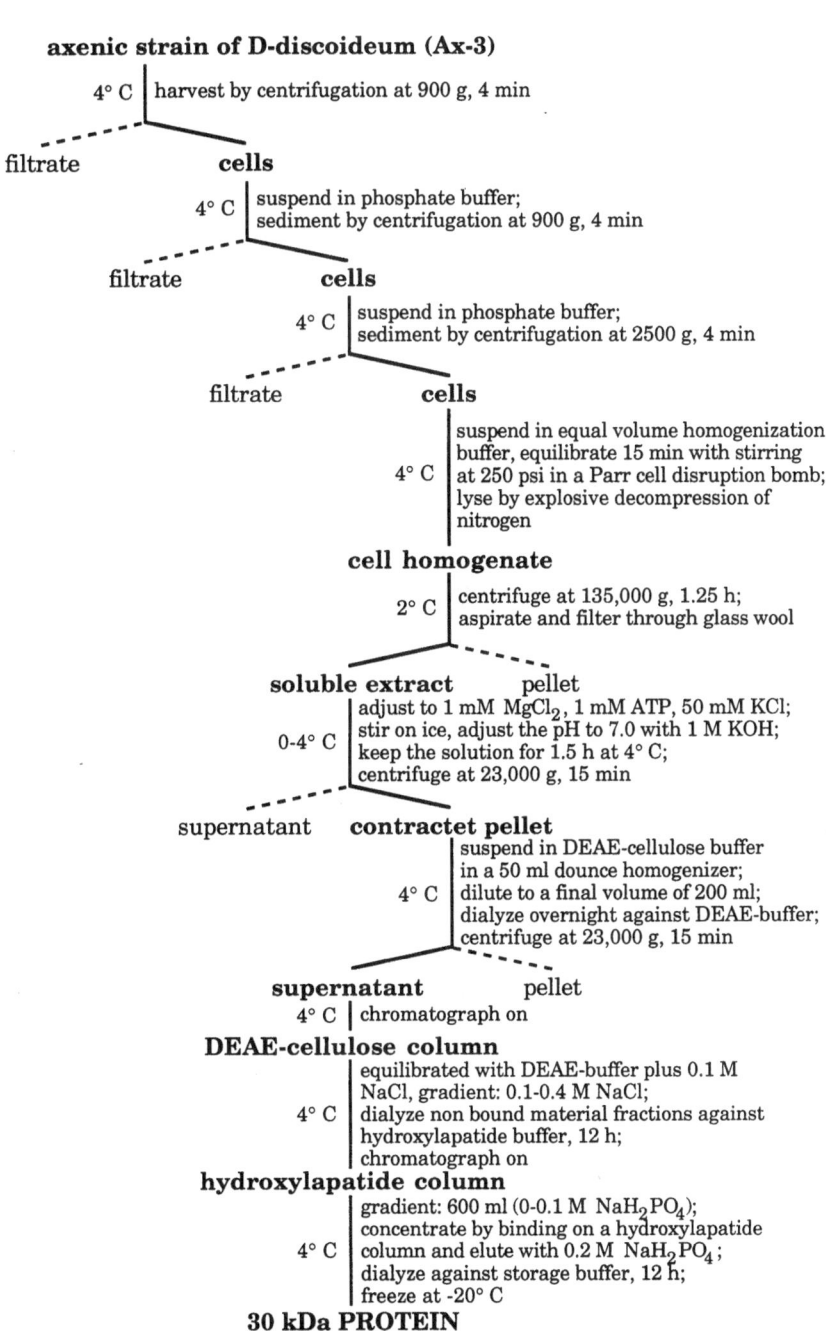

4° C | harvest by centrifugation at 900 g, 4 min

filtrate   **cells**

4° C | suspend in phosphate buffer;
sediment by centrifugation at 900 g, 4 min

filtrate   **cells**

4° C | suspend in phosphate buffer;
sediment by centrifugation at 2500 g, 4 min

filtrate   **cells**

4° C | suspend in equal volume homogenization
buffer, equilibrate 15 min with stirring
at 250 psi in a Parr cell disruption bomb;
lyse by explosive decompression of
nitrogen

cell homogenate

2° C | centrifuge at 135,000 g, 1.25 h;
aspirate and filter through glass wool

soluble extract   pellet

0-4° C | adjust to 1 mM $MgCl_2$, 1 mM ATP, 50 mM KCl;
stir on ice, adjust the pH to 7.0 with 1 M KOH;
keep the solution for 1.5 h at 4° C;
centrifuge at 23,000 g, 15 min

supernatant   **contractet pellet**

4° C | suspend in DEAE-cellulose buffer
in a 50 ml dounce homogenizer;
dilute to a final volume of 200 ml;
dialyze overnight against DEAE-buffer;
centrifuge at 23,000 g, 15 min

supernatant   pellet

4° C | chromatograph on

DEAE-cellulose column

4° C | equilibrated with DEAE-buffer plus 0.1 M
NaCl, gradient: 0.1-0.4 M NaCl;
dialyze non bound material fractions against
hydroxylapatide buffer, 12 h;
chromatograph on

hydroxylapatide column

4° C | gradient: 600 ml (0-0.1 M NaH_2PO_4);
concentrate by binding on a hydroxylapatide
column and elute with 0.2 M NaH_2PO_4;
dialyze against storage buffer, 12 h;
freeze at -20° C

30 kDa PROTEIN
(~0.85 mg/100 ml cells)

ACTIN

The most abundant muscle and non-muscle cytoskeletal protein. MW 42.000 Da, 374/375 amino acids. Various isoforms. Structure resolved at 2.8 Å

Source: rabbit skeletal muscle

Equipment:
- mincer
- glass-Teflon-homogenizer with pestle
- sephacryl S-300 column (5 x 100 cm)
- electrophoretic gels
- centrifuge
- steril cheesecloth

Chemicals: KCl, EDTA, aceton, Trizma Base, $CaCl_2$, $MgCl_2$, ATP, DTT, NaN_3,
column material: sephacryl S-300

Have ready: **buffer G:** 2 mM Tris-HCl, 0.2 mM $CaCl_2$, 0.2 mM ATP, 0.2 mM DTT, 0.05% NaN_3, pH 8.0 (15 liters)

Reference: Sheterline, P. (1994) Actin, Protein Profile Vol. 1, 1, Academic Press
Kabsch, W., Mannherz, H.G., Suck, D., Pai, E.F., and Holmes, K.C. (1990) Nature 347, 37-44

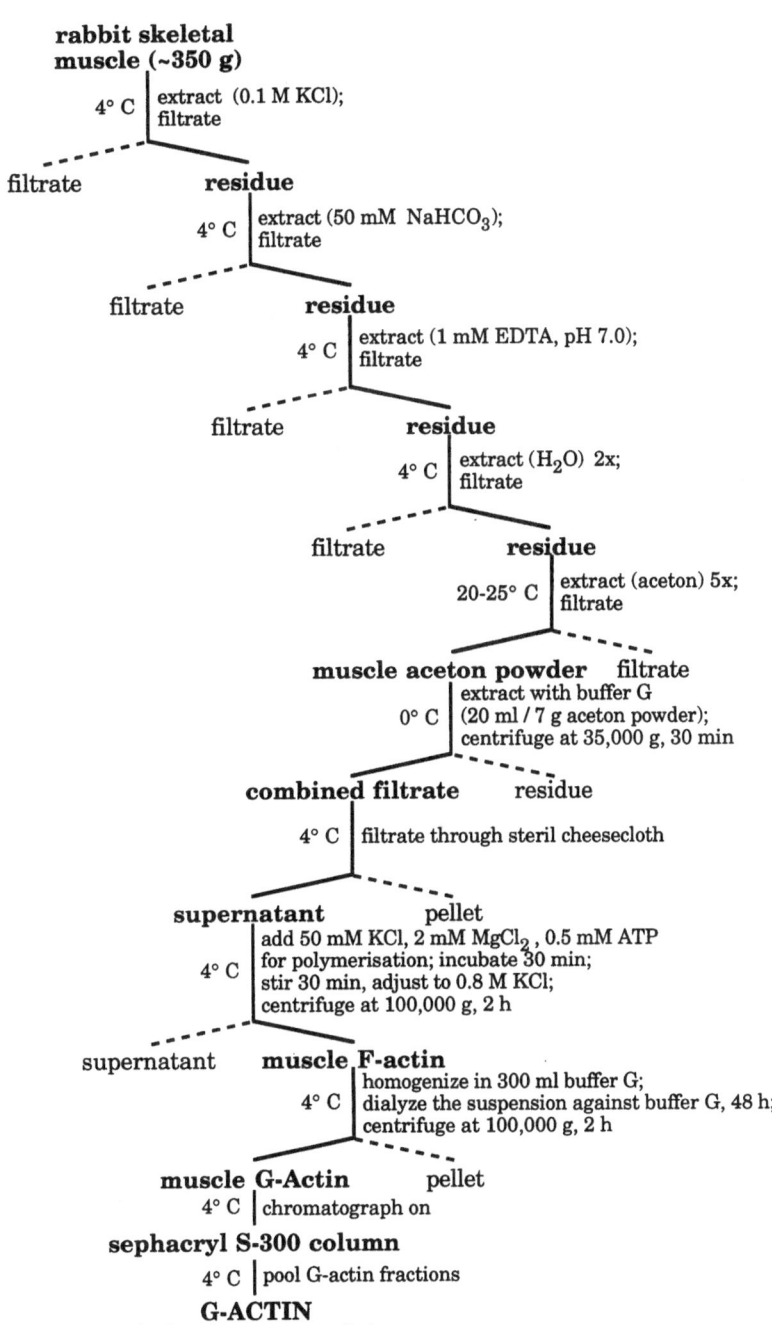

**rabbit skeletal
muscle (~350 g)**

4° C | extract (0.1 M KCl);
filtrate

filtrate   **residue**

4° C | extract (50 mM NaHCO₃);
filtrate

filtrate   **residue**

4° C | extract (1 mM EDTA, pH 7.0);
filtrate

filtrate   **residue**

4° C | extract (H₂O) 2x;
filtrate

filtrate   **residue**

20-25° C | extract (aceton) 5x;
filtrate

muscle aceton powder   filtrate
extract with buffer G
0° C | (20 ml / 7 g aceton powder);
centrifuge at 35,000 g, 30 min

combined filtrate   residue

4° C | filtrate through steril cheesecloth

supernatant   pellet
add 50 mM KCl, 2 mM MgCl₂, 0.5 mM ATP
for polymerisation; incubate 30 min;
4° C | stir 30 min, adjust to 0.8 M KCl;
centrifuge at 100,000 g, 2 h

supernatant   **muscle F-actin**
homogenize in 300 ml buffer G;
4° C | dialyze the suspension against buffer G, 48 h;
centrifuge at 100,000 g, 2 h

muscle G-Actin   pellet
4° C | chromatograph on

sephacryl S-300 column

4° C | pool G-actin fractions

G-ACTIN
(~10 mg/g acetone powder)

ACTIN DEPOLYMERIZING FACTOR (ADF)

Actin depolymerizing factor (ADF) is a 19 kDa protein which severs actin filaments and sequesters actin monomers into a 1:1 complex. Found in a variety of cells.

Source:
a) embryonic chick brain
b) cultured BHK cells
c) cultured myocytes

Equipment:
- Teflon-glass homogenizer
- centrifuge
- DEAE-cellulose column for   a) 5 x 4 cm
       b) 2.5 x 5 cm
       c) 1.5 x 6 cm
- sephadex G-75 column for   a) 5 x 90 cm
       b) 1.5 x 80 cm
       c) 1.5 x 80 cm
- hydroxylapatide column for   b) 1 x 4 cm
       c) 1.5 x 2 cm
- blue B-Sepharose column for   b) 19 kDa
       1 x 2.5 cm
- green A-Sepharose column for
       b) 19 kDa
       1 x 2.5 cm
       b) 20 kDa
       1 x 2.5 cm
       c) 1.5 x 3 cm
- Amicon CF-25 filter cones
- Amicon ultrafiltration cell YM 5 membrane
- culture flask-175 cm^2
- vortex
- clinical centrifuge
- primaria culture dish 100 mm diameter
- sonicator
- electrophoretic gels

Chemicals:
a) Trizma Base, DTT, NaF, NaCl, NaN_3
b) Trizma Base, DTT, NaF, NaCl, NaN_3, DMEM, fetal bovine serum, antibiotic-antimycotic solution, KCl, NaH_2PO_4, trypsin, EDTA
c) Trizma Base, DTT, NaF, NaCl, NaN_3, DMEM, heat-inactivated horse serum, chick embryonic extract, antibiotic-antimycotic solution, NaH_2PO_4
protease inhibitors: tosylarginin-methyl-ester, benzoylarginin-methyl-ester, tosylamide-2-phenylethyl-chloromethyl-ketone, soybean-trypsin

inhibitor, leupeptin, chymostatin, antipain,
pepstatin

Have ready: **buffer A:** 10 mM Tris-HCl, 0.5 mM DTT, pH 7.5

washing buffer: buffer A, plus 50 mM NaCl

extraction buffer: buffer A containing 2 mM NaF,
10 mg/ml tosylarginin-methyl-ester, 10 mg/ml
benzoylarginin-methyl-ester, 10 mg/ml tosylamide-2-
phenylethyl-chloromethyl-ketone, 10 mg/ml soybean-
trypsin inhibitor, 1 mg/ml leupeptin, 1 mg/ml
chymostatin, 1 mg/ml antipain, 1 mg/ml pepstatin

sephadex-column buffer: buffer A, plus 1 mM
NaN$_3$

**PBS (Ca^{2+}/Mg^{2+} free phosphate buffered
saline):** 2.7 mM KCl, 140 mM NaCl, 8 mM
NaH$_2$PO$_4$, pH 7.2

trypsin buffer: PBS-buffer plus 0.25 g/liter
trypsin, 0.2 g/liter EDTA

buffer B: 5 mM NaH$_2$PO$_4$, 0.5 mM DTT, pH 7.5

hydroxylapatide buffer: 10 mM NaH$_2$PO$_4$,
0.5 mM DTT, 10 mM NaF, pH 7.5

Reference: Bamburg, J.R. et al. (1991) Methods Enzymol. 196,
125-140
Gen-Bank: J02912 and J02915

a) brain from 19 day old embryonic chicks

4° C | brains are frozen in liquid nitrogen and stored by -70° C

brain tissue

4° C | homogenize in a Teflon-glass homogenizer with 125 ml buffer A;
centrifuge at 143,000 g, 90 min

supernatant   pellet

4° C | apply on

DEAE-cellulose column

4° C | equilibrated with buffer A;
collect the flow through plus 1 volume of washing buffer;
concentrate with an Amicon CF-25 filter to 7-8 ml;
chromatograph on

sephadex G-25 column

4° C | equilibrated with sephadex column buffer;
pool the ADF fraction
chromatograph on

green A sepharose column

| gradient: 250 ml (0-150 mM NaCl)

ADF

b) baby hamster kidney (BHK-21/C13)

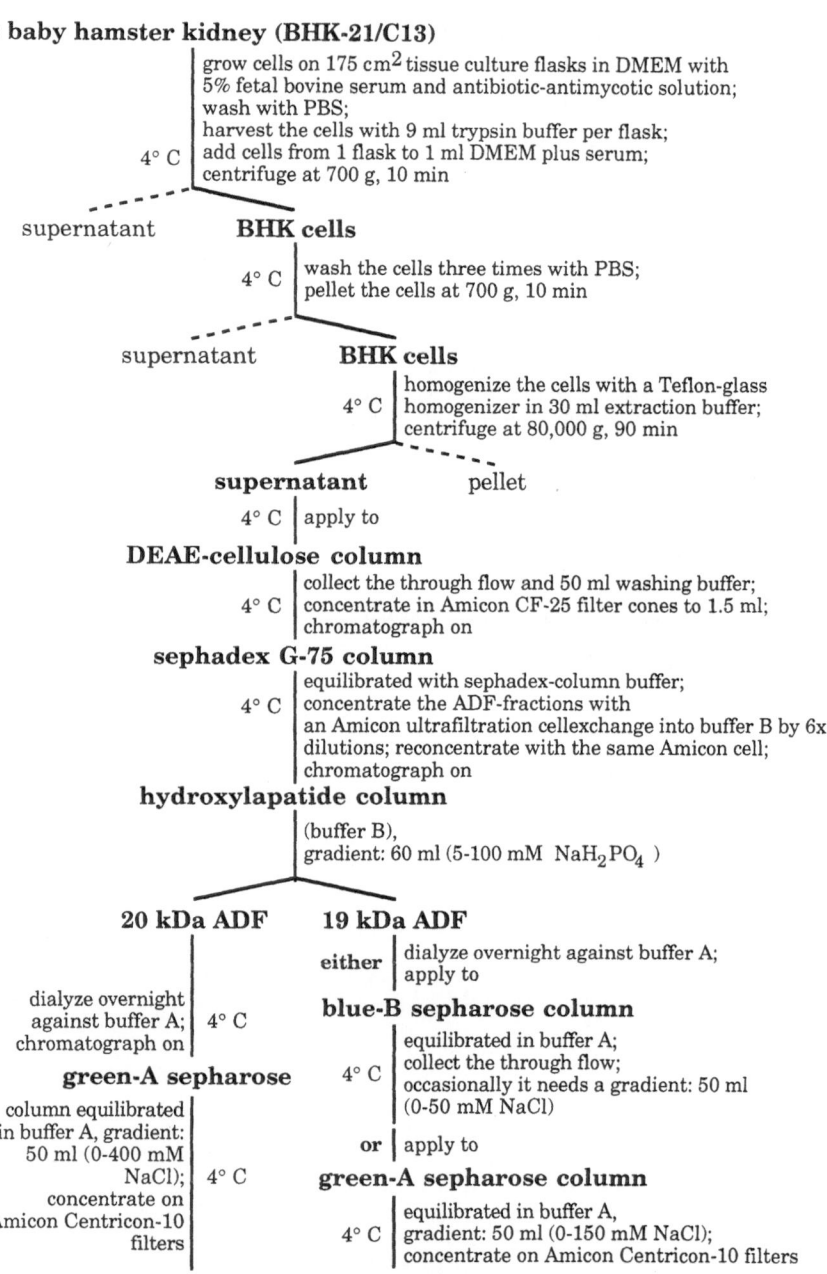

4° C | grow cells on 175 cm^2 tissue culture flasks in DMEM with 5% fetal bovine serum and antibiotic-antimycotic solution; wash with PBS; harvest the cells with 9 ml trypsin buffer per flask; add cells from 1 flask to 1 ml DMEM plus serum; centrifuge at 700 g, 10 min

supernatant   **BHK cells**

4° C | wash the cells three times with PBS; pellet the cells at 700 g, 10 min

supernatant   **BHK cells**

4° C | homogenize the cells with a Teflon-glass homogenizer in 30 ml extraction buffer; centrifuge at 80,000 g, 90 min

supernatant   pellet

4° C | apply to

DEAE-cellulose column

4° C | collect the through flow and 50 ml washing buffer; concentrate in Amicon CF-25 filter cones to 1.5 ml; chromatograph on

sephadex G-75 column

4° C | equilibrated with sephadex-column buffer; concentrate the ADF-fractions with an Amicon ultrafiltration cellexchange into buffer B by 6x dilutions; reconcentrate with the same Amicon cell; chromatograph on

hydroxylapatide column

(buffer B), gradient: 60 ml (5-100 mM NaH$_2$PO$_4$)

20 kDa ADF   **19 kDa ADF**

either | dialyze overnight against buffer A; apply to

dialyze overnight against buffer A; chromatograph on | 4° C

blue-B sepharose column

4° C | equilibrated in buffer A; collect the through flow; occasionally it needs a gradient: 50 ml (0-50 mM NaCl)

green-A sepharose

column equilibrated in buffer A, gradient: 50 ml (0-400 mM NaCl); concentrate on Amicon Centricon-10 filters | 4° C

or | apply to

green-A sepharose column

4° C | equilibrated in buffer A, gradient: 50 ml (0-150 mM NaCl); concentrate on Amicon Centricon-10 filters

20 kDa ADF   **19 kDa ADF**

c) pectoral muscle from 30
11 day old embryonic chicks

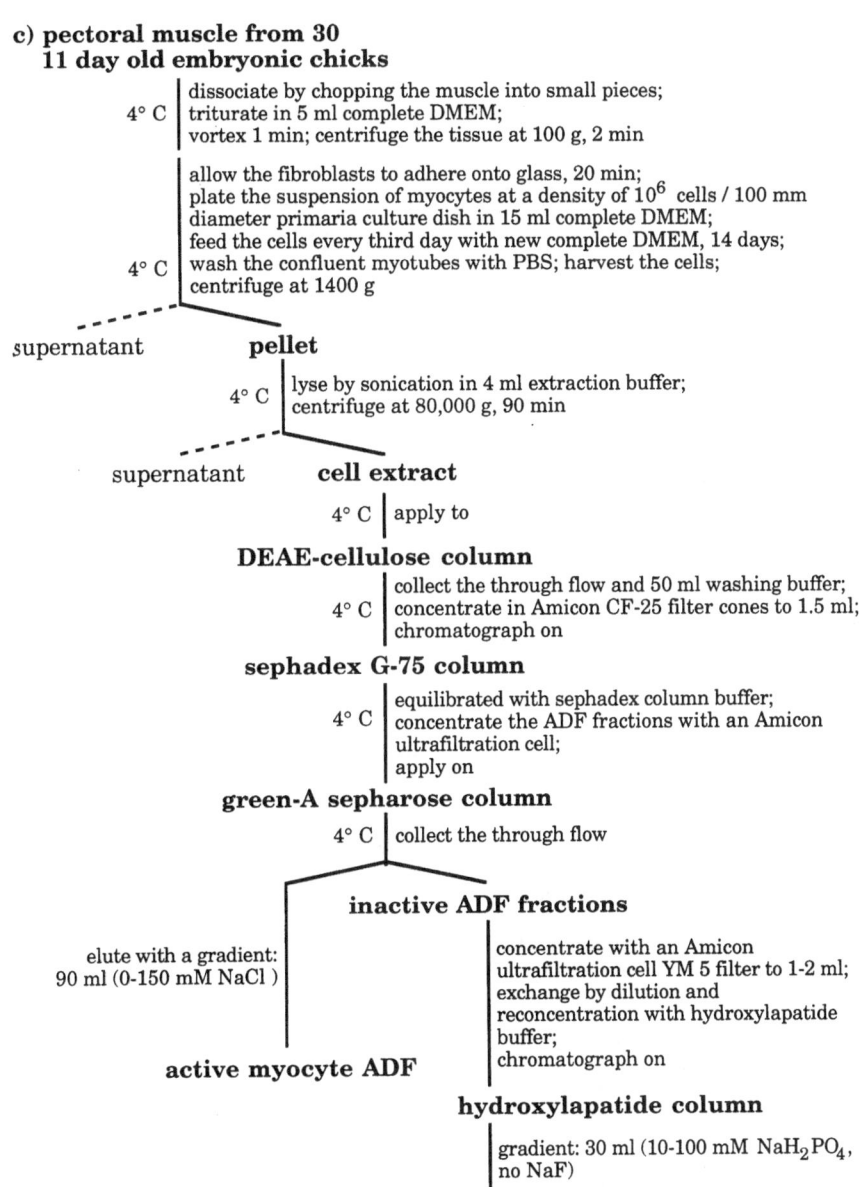

	dissociate by chopping the muscle into small pieces;
4° C	triturate in 5 ml complete DMEM;
	vortex 1 min; centrifuge the tissue at 100 g, 2 min

	allow the fibroblasts to adhere onto glass, 20 min;
	plate the suspension of myocytes at a density of 10^6 cells / 100 mm
	diameter primaria culture dish in 15 ml complete DMEM;
	feed the cells every third day with new complete DMEM, 14 days;
4° C	wash the confluent myotubes with PBS; harvest the cells;
	centrifuge at 1400 g

supernatant   **pellet**

| | lyse by sonication in 4 ml extraction buffer; |
| 4° C | centrifuge at 80,000 g, 90 min |

supernatant   **cell extract**

4° C | apply to

DEAE-cellulose column

	collect the through flow and 50 ml washing buffer;
4° C	concentrate in Amicon CF-25 filter cones to 1.5 ml;
	chromatograph on

sephadex G-75 column

	equilibrated with sephadex column buffer;
4° C	concentrate the ADF fractions with an Amicon
	ultrafiltration cell;
	apply on

green-A sepharose column

4° C | collect the through flow

inactive ADF fractions

elute with a gradient:
90 ml (0-150 mM NaCl)

concentrate with an Amicon
ultrafiltration cell YM 5 filter to 1-2 ml;
exchange by dilution and
reconcentration with hydroxylapatide
buffer;
chromatograph on

active myocyte ADF

hydroxylapatide column

gradient: 30 ml (10-100 mM NaH_2PO_4,
no NaF)

inactive myocyte ADF

α-ACTININ

A 94-103 kDa F-actin crosslinking protein from muscle and non-muscle cells forming antiparallel homodimers of 30-40 nm in length.

Source: chicken or turkey gizzard smooth muscle

Equipment:
- mincer
- sharp knife
- DEAE-cellulose column (2.5 x 16 cm)
- sepharose 4B column (2.5 x 90 cm)
- hydroxylapatide column (1.6 x 10 cm)
- Amicon cell
- electrophoretic gels
- centrifuge

Chemicals: glycerine (anhydrous, extra pure), $KHCO_3$, KCl, $(NH_4)_2SO_4$ (up), Trizma Base, EDTA, DTT, $MgCl_2$, NaN_3, KH_2PO_4

protease inhibitors: trasylol (aprotinin)

Have ready: **extraction buffer:** 1 mM $KHCO_3$+10 U of trasylol /ml, pH 7.4-7.5 (36 liters)

DEAE-column buffer: 10 mM Tris-HCl, 1 mM EDTA, 1 mM DTT, 2 mM $MgCl_2$, 10 U of trasylol /ml, pH 7.4 (4 liters)

sepharose-column buffer: 10 mM Tris-HCl, 0.6 M KCl, 1 mM DTT, 1 mM EDTA, 2 mM $MgCl_2$, 0.02% NaN_3, pH 7.4 (1 liter)

hydroxylapatide-column buffer: 50 mM KH_2PO_4, 0.1 mM EDTA, pH 7.0 (5 liters)

storage buffer: 1 mM $KHCO_3$, 0.02% NaN_3, pH 7.0-8.0

Reference: Craig, S.W., Lancashire, Ch.L. and Cooper, J.A. (1982) Methods Enzymol. 85, 316-321
Gen-Bank: X51753 (chicken), J03486 (smooth muscle), X15804 (non-muscle), Y00689 (Dictyostelium)

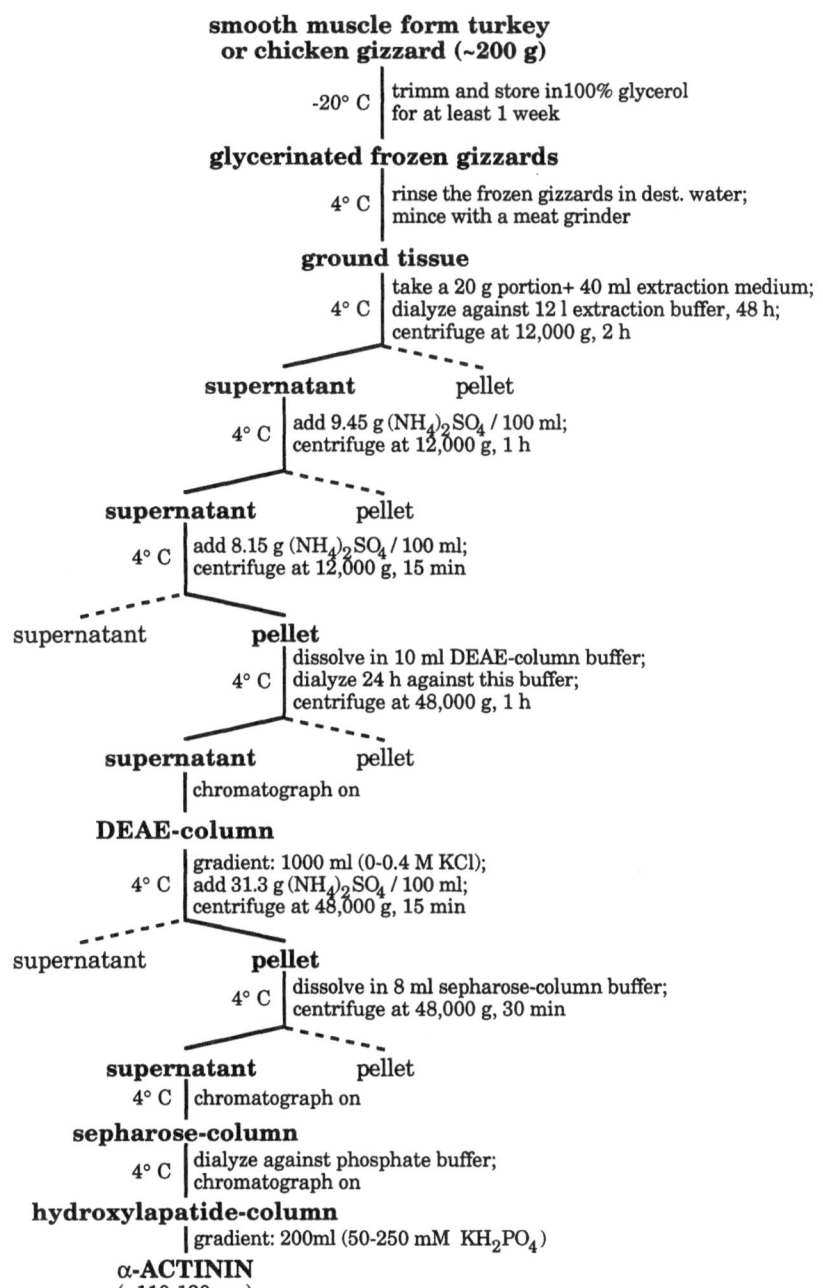

**smooth muscle form turkey
or chicken gizzard (~200 g)**

-20° C | trimm and store in100% glycerol
for at least 1 week

glycerinated frozen gizzards

4° C | rinse the frozen gizzards in dest. water;
mince with a meat grinder

ground tissue

4° C | take a 20 g portion+ 40 ml extraction medium;
dialyze against 12 l extraction buffer, 48 h;
centrifuge at 12,000 g, 2 h

supernatant   pellet

4° C | add 9.45 g $(NH_4)_2SO_4$ / 100 ml;
centrifuge at 12,000 g, 1 h

supernatant   pellet

4° C | add 8.15 g $(NH_4)_2SO_4$ / 100 ml;
centrifuge at 12,000 g, 15 min

supernatant   **pellet**

4° C | dissolve in 10 ml DEAE-column buffer;
dialyze 24 h against this buffer;
centrifuge at 48,000 g, 1 h

supernatant   pellet

| chromatograph on

DEAE-column

4° C | gradient: 1000 ml (0-0.4 M KCl);
add 31.3 g $(NH_4)_2SO_4$ / 100 ml;
centrifuge at 48,000 g, 15 min

supernatant   **pellet**

4° C | dissolve in 8 ml sepharose-column buffer;
centrifuge at 48,000 g, 30 min

supernatant   pellet

4° C | chromatograph on

sepharose-column

4° C | dialyze against phosphate buffer;
chromatograph on

hydroxylapatide-column

| gradient: 200ml (50-250 mM KH_2PO_4)

α-ACTININ
(~110-130 mg)

ACTOBINDIN

A 9.7 kDa protein with 88 amino acids which binds to monomeric actin and actin nuclei thus inhibiting actin polymerization.

Source: acanthamoeba castellanii

Equipment:
- carboy 15 liters
- centrifuge
- dounce tissue grinder 100 ml
- chemical hood
- DEAE-cellulose column (5 x 60 cm)
- hydroxylapatide column (70 ml)
- sephacryl HR-100 column (1.6 x 95 cm)
- low molecular weight cutoff dialysis tubing
- cutoff collodion bag M_r 10,000
- *occasionally:* polyproline affinity column
- electrophoretic gels

Chemicals: Trizma Base, NaCl, $(NH_4)_2SO_4$, KH_2PO_4, NaN_3
protease inhibitors: leupeptin, pepstatin, PMSF, DFP

Have ready: **washing buffer:** 10 mM Tris-HCl, 200 mM NaCl, pH 8.0

homogenization buffer: 10 mM Tris-HCl, 600 mM NaCl, 1 mg/liter leupeptin, 1 mg/liter pepstatin, 75 mg/liter PMSF, 100 mM DFP, pH 8.0

DEAE-column buffer: 5 mM Tris-HCl, pH 8.0

hydroxylapatide-column buffer: 5 mM KH_2PO_4, pH 6.7

sephacryl-column buffer: 5 mM Tris-HCl, 0.04% (w/v) NaN_3, pH 8.0

Reference: Bubb, M.R. and Korn, E.D. (1990) Methods Enzymol. 196, 119-125

Acanthamoebe castellanii

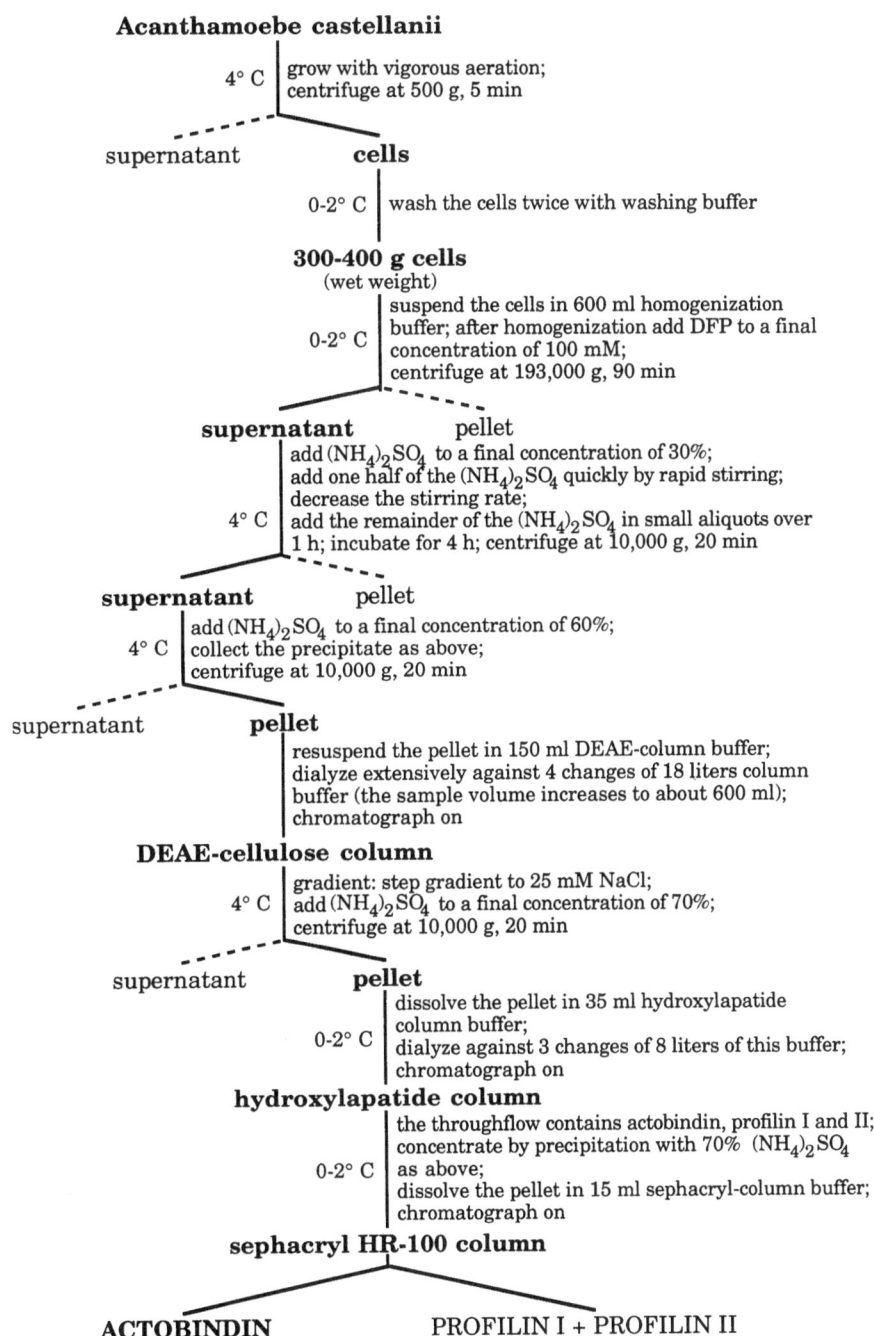

4° C | grow with vigorous aeration;
centrifuge at 500 g, 5 min

- - - - supernatant   **cells**

0-2° C | wash the cells twice with washing buffer

300-400 g cells
(wet weight)

0-2° C | suspend the cells in 600 ml homogenization buffer; after homogenization add DFP to a final concentration of 100 mM; centrifuge at 193,000 g, 90 min

supernatant   pellet

4° C | add $(NH_4)_2SO_4$ to a final concentration of 30%; add one half of the $(NH_4)_2SO_4$ quickly by rapid stirring; decrease the stirring rate; add the remainder of the $(NH_4)_2SO_4$ in small aliquots over 1 h; incubate for 4 h; centrifuge at 10,000 g, 20 min

supernatant   pellet

4° C | add $(NH_4)_2SO_4$ to a final concentration of 60%; collect the precipitate as above; centrifuge at 10,000 g, 20 min

- - - - supernatant   **pellet**

resuspend the pellet in 150 ml DEAE-column buffer; dialyze extensively against 4 changes of 18 liters column buffer (the sample volume increases to about 600 ml); chromatograph on

DEAE-cellulose column

4° C | gradient: step gradient to 25 mM NaCl; add $(NH_4)_2SO_4$ to a final concentration of 70%; centrifuge at 10,000 g, 20 min

- - - - supernatant   **pellet**

0-2° C | dissolve the pellet in 35 ml hydroxylapatide column buffer; dialyze against 3 changes of 8 liters of this buffer; chromatograph on

hydroxylapatide column

0-2° C | the throughflow contains actobindin, profilin I and II; concentrate by precipitation with 70% $(NH_4)_2SO_4$ as above; dissolve the pellet in 15 ml sephacryl-column buffer; chromatograph on

sephacryl HR-100 column

ACTOBINDIN   PROFILIN I + PROFILIN II

ACTOLINKIN

A 20 kDa monomeric protein binding to actin monomers and free barbed actin filament ends

Source:	echinoderm eggs
Equipment:	• centrifuge
	• sephacryl S-300 column (2.6 x 90 cm)
	• DNase I-affinity-gel 10 column
	• hydroxylapatide column (2ml)
	• sephacryl S-300 column (1.6 x 60 cm)
	• sephadex G-100 column (1.6 x 30 cm)
	• electrophoretic gels
Chemicals:	KCl, $MgCl_2$, EGTA, MOPS, DTT, NaOH, ATP, urea, KH_2PO_4, aquacide II-A powder
	protease inhibitors: PMSF, leupeptin
Have ready:	**extraction buffer I:** 0.1 M KCl, 2 mM $MgCl_2$, 5 mM EGTA, 1 mM DTT, 1 mM ATP, 1 mM PMSF, 10 mg/ml leupeptin, 10 mM MOPS-NaOH, pH 6.8
	extraction buffer II: extraction buffer I adjusted to 1.1 M KCl
	KCl buffer: 0.3 M KCl, 10 mM MOPS-NaOH, pH 7.2
	elution buffer: 4 M urea, 10 mM MOPS-NaOH, 0.5 mM DTT, pH 7.2
	hydroxylapatide buffer: 10 mM KH_2PO_4, pH 6.8
	dialysis buffer: aquacide II-A powder, 50 mM KCl, 2 mM $MgCl_2$, 0.5 mM DTT, 1 mM NaN_3, 10 mM MOPS-NaOH 1 mg/ml leupeptin, pH 7.2
	sephacryl S-300 buffer: 7 M urea, 10 mM MOPS-NaOH, 0.5 mM DTT, 5 mg/ml leupeptin, pH 7.2
	sephadex G-100 buffer: 50 mM KCl, 10 mM MOPS-NaOH, pH 7.2
Reference:	Ishidate, S. and Mabuchi, I. (1988) J. Biochem. 104, 72-80

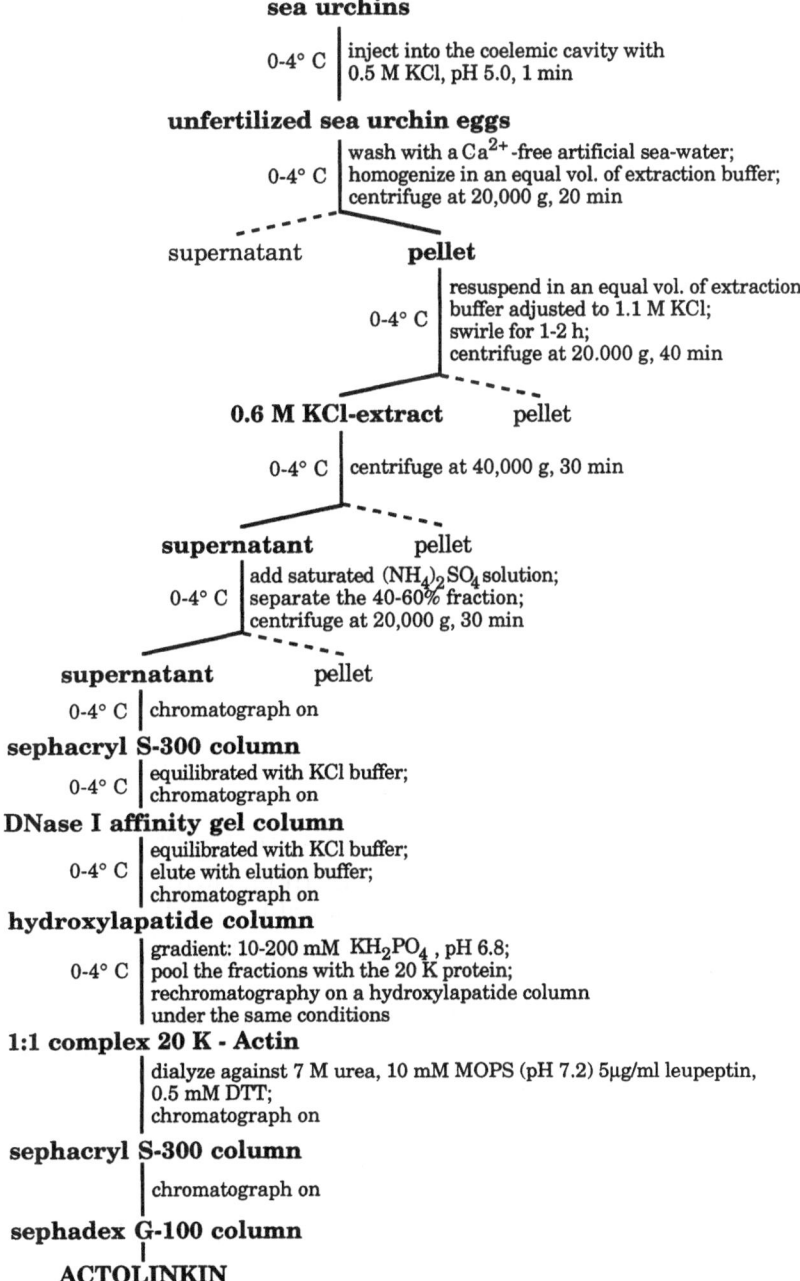

sea urchins

0-4° C | inject into the coelemic cavity with 0.5 M KCl, pH 5.0, 1 min

unfertilized sea urchin eggs

0-4° C | wash with a Ca^{2+}-free artificial sea-water; homogenize in an equal vol. of extraction buffer; centrifuge at 20,000 g, 20 min

supernatant   **pellet**

0-4° C | resuspend in an equal vol. of extraction buffer adjusted to 1.1 M KCl; swirl for 1-2 h; centrifuge at 20.000 g, 40 min

0.6 M KCl-extract   pellet

0-4° C | centrifuge at 40,000 g, 30 min

supernatant   pellet

0-4° C | add saturated $(NH_4)_2SO_4$ solution; separate the 40-60% fraction; centrifuge at 20,000 g, 30 min

supernatant   pellet

0-4° C | chromatograph on

sephacryl S-300 column

0-4° C | equilibrated with KCl buffer; chromatograph on

DNase I affinity gel column

0-4° C | equilibrated with KCl buffer; elute with elution buffer; chromatograph on

hydroxylapatide column

0-4° C | gradient: 10-200 mM KH_2PO_4, pH 6.8; pool the fractions with the 20 K protein; rechromatography on a hydroxylapatide column under the same conditions

1:1 complex 20 K - Actin

| dialyze against 7 M urea, 10 mM MOPS (pH 7.2) 5µg/ml leupeptin, 0.5 mM DTT; chromatograph on

sephacryl S-300 column

| chromatograph on

sephadex G-100 column

ACTOLINKIN
(~0.1 mg / 200 ml eggs)

ADSEVERIN

A 74 kDa monomeric barbed-end F-actin capping and severing protein which is Ca^{++}-activated and PIP-2, PI and PS-inhibited.

Source:	bovine adrenal medulla
Equipment:	• centrifuge • DNA-se I-Affi gel 10 affinity column (4 x 12 cm) (cf. DESTRIN prep.) • hydroxylapatide column (1.6 x 6 cm) • DEAE-cellulose column (DE-52) (1.3 x 4 cm) • HPLC-gel filtration column (S 3000 GW, Toyo Soda) (0.8 x 60 cm)
Chemicals:	Tris-HCl, EGTA, ATP, DTT, HEPES-KOH, NaCl, MgCl$_2$, CaCl$_2$, K/PO$_4$, MES-KOH, EDTA, guanidine-HCl **protease inhibitors:** PMSF
Have ready:	**homogenization solution:** 10 mM Tris-HCl, 1 mM EGTA, 0.2 mM DTT, 0.2 mM ATP, 1 mM PMSF, pH 7.8
	buffer A: 2 mM HEPES-KOH, 0.1 mM ATP, 0.1 mM DTT, pH 7.2
	buffer D: 10 mM HEPES-KOH, 1 mM MgCl$_2$, 1 mM CaCl$_2$, 0.1 mM DTT, pH 7.2
	elution buffer I: buffer D + 0.6 M NaCl
	elution buffer II: 10 mM HEPES-KOH, 10 mM EGTA, 1 mM MgCl$_2$, 1 M NaCl, 0.1 mM DTT, pH 7.2
	elution buffer III: 10 mM MES-KOH, 2 mM EDTA, 3 M guanidine-HCl, pH 6.8
	dialyzing buffer: 10 mM Tris-HCl, 0.2 mM EGTA, 0.1 mM DTT, pH 7.8
	HT-column buffer: 20 mM K/PO$_4$, 50 mM NaCl, pH 7.0
	DEAE-column buffer: 10 mM Tris-HCl, 0.2 mM EGTA, 0.1 mM DTT, pH 7.8
	HPLC-gel filtration buffer: 10 mM MES-KOH, 0.8 M NaCl, 0.2 mM EGTA, 0.1 mM DTT, pH 6.8
Reference:	Maekawa, S. et al. (1989) J. Biol. Chem. 264, 7458-7465 Maekawa, S. and Sakai, H. (1990) J. Biol. Chem. 265, 10940-10942

adrenal medulla (~100 g)

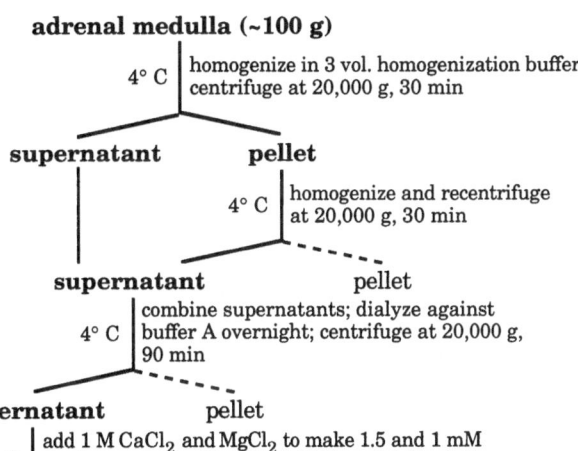

4° C | homogenize in 3 vol. homogenization buffer;
centrifuge at 20,000 g, 30 min

supernatant   **pellet**

4° C | homogenize and recentrifuge
at 20,000 g, 30 min

supernatant   pellet

4° C | combine supernatants; dialyze against
buffer A overnight; centrifuge at 20,000 g,
90 min

supernatant   pellet

4° C | add 1 M $CaCl_2$ and $MgCl_2$ to make 1.5 and 1 mM
final conc.; mix with 150 ml DNA-se I-Affi gel-10
resin equilibrated in buffer D, 60 min; pour

DNA-se I-Affi gel affinity column

4° C | wash with buffer D; step elute by elution buffer I, II and III;
pool eluate II; dialyze against dialyzing buffer;
apply to

hydroxylapatide column

4° C | preequilibrate with HT-column buffer; wash and elute
with linear gradient (50-350 mM phosphate);
collect fractions into 200 mM EGTA to give a final
conc. of 2 mM EGTA; pool 74 kDa protein (third
protein peak); apply to

DEAE-cellulose column

4° C | preequilibrated with DEAE column buffer; elute
with linear gradient (0-300 mM NaCl);
pool 74 kDa protein; apply to

HPLC-gel filtration column

4° C | elute with HPLC gel filtration
buffer

ADSEVERIN (~50 μg)

AGINACTIN

A globular, monomeric, barbed-end F-actin capping protein of 70 kDa molecular weight. A Ca^{++} independent, non severing and non-nucleating but agonist-regulated actin binding protein.

Source: Dictyostelium discoideum

Equipment:
- DEAE-cellulose (DE-52) column (5 x 26 cm)
- phenyl-sepharose column (5 x 20 cm) Pharmacia
- DEAE-500 (Perkin Elmer) ion exchange column (1 x 15 cm)
- FPLC-system (Pharmacia)
- hydroxylapatide (Bio gel, BioRad) column (1 x 8 cm)
- Mono Q-HR 5/5 column (Pharmacia)
- phospho-cellulose column, P-11 (Whatman) (1 x 10 cm)

Chemicals: KCl, PIPES, EGTA, $MgCl_2$, DTT, ATP, Triton X-100, NaCl, $(NH_4)_2SO_4$, KH_2PO_4, K/PO_4, Na-pyrophosphate, EDTA, triethanolamine, NaN_3 **protease inhibitors:** leupeptin, pepstatin A, chymostatin

Have ready: **lysis buffer:** 140 mM KCl, 80 mM PIPES, 20 mM EGTA, 0.4 mM $MgCl_2$, 20 mM DTT, 4 mM ATP, 2% Triton X-100, pH 7.0 + 5 µg/ml each leupeptin, pepstatin A and chymostatin

DE-52 buffer: 35 mM KCl, 20 mM PIPES, 0.5 mM EGTA, 0.1 mM $MgCl_2$, 1 mM DTT, 0.5 mM ATP, pH 7.0

PS-buffer: 10 mM KH_2PO_4, 4 mM Na-pyrophosphate, 2 mM EDTA, 0.5 mM DTT, pH 7.0 + indicated $(NH_4)_2SO_4$ conc.

dialyzing buffer: 20 mM KCl, 5 mM PIPES, 1.25 mM EGTA, 0.05 mM $MgCl_2$, 0.5 mM DTT, pH 7.0

PE-buffer: 2 mM triethanolamine, 0.5 mM DTT, pH 7.5

MTP-buffer: 50 mM KCl, 10 mM PIPES, 5 mM EDTA, 0.25 mM DTT, 0.02% NaN_3, pH 7.0

assay buffer: 50 mM KCl, 10 mM PIPES, 2 mM EGTA, 1 mM $MgCl_2$, 0.5 mM DTT, 0.5 mM ATP, 0.02% NaN_3, pH 7.0

Mono Q-buffer: 20 mM triethanolamine, 0.1 mM DTT, 0.1 mM EGTA, pH 7.3

PC-buffer: 20 mM PIPES, 2 mM EGTA, 0.1 mM DTT, 0.02% NaN_3, pH 6.9

Reference: Sauterer, R.A. et al. (1991) J. Biol. Chem 266, 24533-24539

Dictyostelium cells (1×10^{11} cells/l)

$22°$ C — add 1/3 vol. lysis buffer;
stir at $22°$ C; chill on ice;
$0\text{-}4°$ C — clarify at 16,000 g, 10 min in GSA-rotor

supernatant   pellet

load onto

DEAE-cellulose column

$4°$ C — preequilibrate with DE-52 buffer; elute with 2 x 1000 ml
linear gradient of 35-250 mM NaCl in DE-52 buffer at
150 ml/h; collect 20 ml fractions; pool fractions between
35-130 mM NaCl; make 35% $(NH_4)_2SO_4$; equilibrate at
$0°$ C — $0°$ C, 60 min; clarify at 19,000 g, 70 min in GSA-rotor

supernatant   pellet

load onto

phenyl-sepharose column

preequilibrated with DE-52 buffer; elute with PS-buffer
(150 ml/h); collect 20 ml fractions; step elute with 800 ml
17.5% $(NH_4)_2SO_4$ in PS-buffer; follow with 1 l linear gradient
17.5%-0% $(NH_4)_2SO_4$ in PS-buffer; pool fractions between
6.1-1.1% $(NH_4)_2SO_4$; dialyze against dialyzing buffer
(5 x 1 h, 4500 ml);
load onto

DEAE-500 column

apply a 10 ml linear gradient 0-50 mM NaCl in PE-buffer;
wash with 20 ml 50 mM NaCl in PE-buffer; elute with linear gradient
50-300 mM NaCl in PE-buffer; collect 2 ml fractions; pool 95-160 mM
NaCl eluate; dilute with equal vol. HTP-buffer;
load onto

hydroxylapatide column

preequilibrate with column buffer; elute with 100 ml linear gradient
0-200 mM K/PO_4 in HTP-buffer at 20 ml/h; collect 2 ml fractions; pool
85-120 mM phosphate eluate; dialyze against assay buffer

load onto   dialyze against
                             PC-buffer and load onto

Mono-Q column   **phosphocellulose column**

preequilibrated with   preequilibrate with
Mono-Q buffer;   PC-buffer;
elute with linear gradient   elute with
0-250 mM NaCl   0-250 mM NaCl

AGINACTIN   **AGINACTIN**

ANNEXINS (Calcimedins)

A family of 35 kDa Ca^{++}-dependent actin and phospholipid binding proteins. Probably involved in mediating intracellular Ca^{++}-signals.

Source: rat liver

Equipment:
- centrifuge
- tissue mixer
- glass wool
- phenyl-sepharose column (~1 l resin)
- DE-52 (Whatman) column (2.5 x 18 cm)
- Ultrogel AcA 44 gel filtration column (1 x 100 cm)

Chemicals: Tris-HCl, EDTA, NaN_3, 2-mercaptoethanol, NaCl, EGTA, Mg-acetate, imidazole, Ponceau S, $CaCl_2$
protease inhibitors: PMSF

Have ready: **homogenization buffer:** 20 mM Tris-HCl, 5 mM EDTA, 7 mM 2-mercaptoethanol, 400 mM NaCl, 0.02% NaN_3, pH 7.4 + 1 mM PMSF

phenyl-Sepharose column buffer: 20 mM Tris-HCl, 1 mM EGTA, 1 mM Mg-acetate, 7 mM 2-mercaptoethanol, 0.02% NaN_3, 200 mM NaCl, pH 7.4

imidazole buffer: 10 mM imidazole, pH 6.2

washing buffer: 20 mM imidazole, 1 mM EGTA, 1 mM Mg-acetate, 7 mM 2-mercaptoethanol, 0.02% NaN_3, pH 6.2

Reference: Kaetzel, M.A. et al. (1989) J. Biol. Chem. 264, 14463-14470

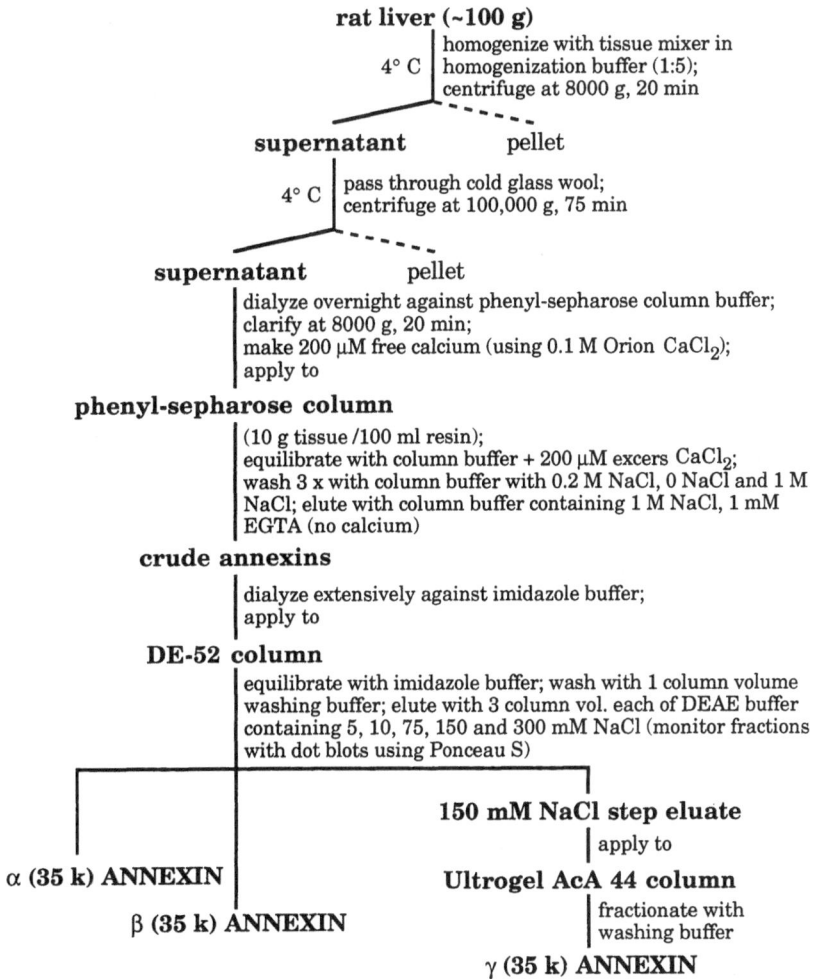

rat liver (~100 g)

4° C | homogenize with tissue mixer in homogenization buffer (1:5); centrifuge at 8000 g, 20 min

supernatant   pellet

4° C | pass through cold glass wool; centrifuge at 100,000 g, 75 min

supernatant   pellet

dialyze overnight against phenyl-sepharose column buffer;
clarify at 8000 g, 20 min;
make 200 µM free calcium (using 0.1 M Orion CaCl$_2$);
apply to

phenyl-sepharose column

(10 g tissue /100 ml resin);
equilibrate with column buffer + 200 µM excers CaCl$_2$;
wash 3 x with column buffer with 0.2 M NaCl, 0 NaCl and 1 M
NaCl; elute with column buffer containing 1 M NaCl, 1 mM
EGTA (no calcium)

crude annexins

dialyze extensively against imidazole buffer;
apply to

DE-52 column

equilibrate with imidazole buffer; wash with 1 column volume
washing buffer; elute with 3 column vol. each of DEAE buffer
containing 5, 10, 75, 150 and 300 mM NaCl (monitor fractions
with dot blots using Ponceau S)

α (35 k) ANNEXIN

β (35 k) ANNEXIN

150 mM NaCl step eluate

| apply to

Ultrogel AcA 44 column

fractionate with
washing buffer

γ (35 k) ANNEXIN

49

C-PROTEIN

A 140-150 kDa myosin and actin binding protein predominantly located in the A-band of skeletal muscle. A rod-shaped molecule of 3 x 35-40 nm, devoid of α-helices.

Source:	rabbit skeletal muscle
Equipment:	• centrifuge • DEAE-sephadex column (5 x 30 cm) • hydroxylapatide column (2.5 x 30 cm)
Chemicals:	KCl, K_2HPO_4, KH_2PO_4, EDTA, $(NH_4)_2SO_4$
Have ready:	**extraction solution:** 0.3 M KCl, 0.05 M K_2HPO_4, 0.1 M KH_2PO_4
	buffer C: 0.5 M KCl, 32.5 mM K_2HPO_4, 17.5 mM KH_2PO_4, 1 mM EDTA, pH 7.0
	buffer D: 135 mM K_2HPO_4, 15.3 mM KH_2PO_4, 10 mM EDTA, pH 7.6
	buffer E: 0.3 M KCl, 4.8 mM K_2HPO_4, 5.2 mM KH_2PO_4, pH 7.0
Reference:	Starr, R. and Offer, G. (1982) Methods Enzymol. 85, 130-138 Gen-Bank: M31209 (chicken)

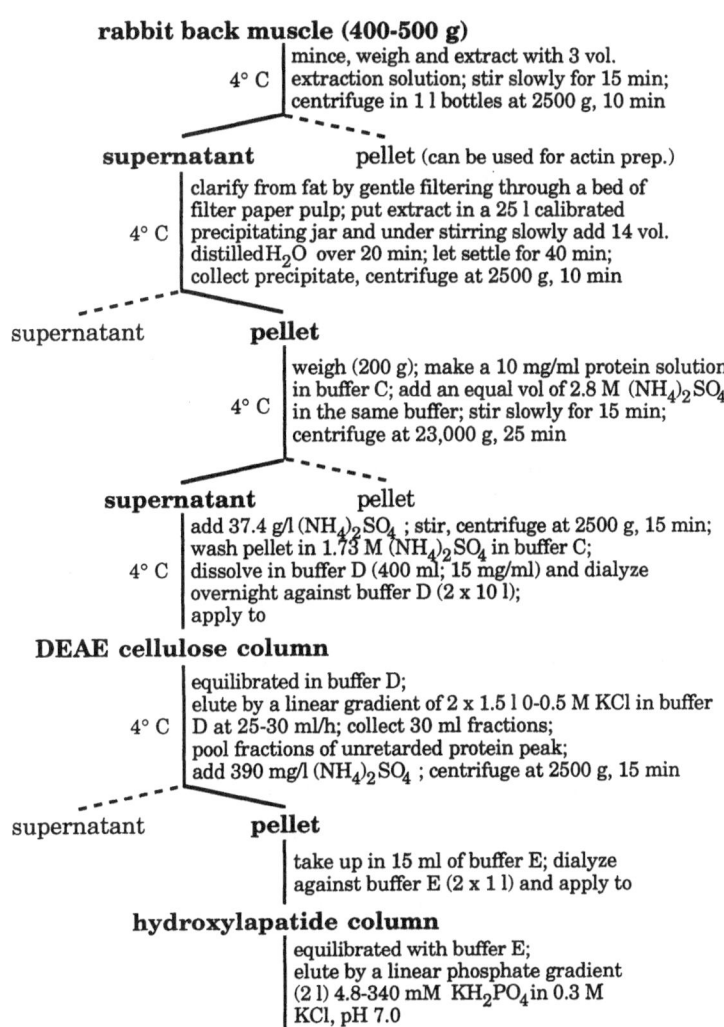

rabbit back muscle (400-500 g)

4° C │ mince, weigh and extract with 3 vol.
     │ extraction solution; stir slowly for 15 min;
     │ centrifuge in 1 l bottles at 2500 g, 10 min

supernatant   pellet (can be used for actin prep.)

4° C │ clarify from fat by gentle filtering through a bed of
     │ filter paper pulp; put extract in a 25 l calibrated
     │ precipitating jar and under stirring slowly add 14 vol.
     │ distilled H_2O over 20 min; let settle for 40 min;
     │ collect precipitate, centrifuge at 2500 g, 10 min

supernatant   **pellet**

4° C │ weigh (200 g); make a 10 mg/ml protein solution
     │ in buffer C; add an equal vol of 2.8 M $(NH_4)_2SO_4$
     │ in the same buffer; stir slowly for 15 min;
     │ centrifuge at 23,000 g, 25 min

supernatant   pellet

4° C │ add 37.4 g/l $(NH_4)_2SO_4$; stir, centrifuge at 2500 g, 15 min;
     │ wash pellet in 1.73 M $(NH_4)_2SO_4$ in buffer C;
     │ dissolve in buffer D (400 ml; 15 mg/ml) and dialyze
     │ overnight against buffer D (2 x 10 l);
     │ apply to

DEAE cellulose column

4° C │ equilibrated in buffer D;
     │ elute by a linear gradient of 2 x 1.5 l 0-0.5 M KCl in buffer
     │ D at 25-30 ml/h; collect 30 ml fractions;
     │ pool fractions of unretarded protein peak;
     │ add 390 mg/l $(NH_4)_2SO_4$; centrifuge at 2500 g, 15 min

supernatant   **pellet**

     │ take up in 15 ml of buffer E; dialyze
     │ against buffer E (2 x 1 l) and apply to

hydroxylapatide column

     │ equilibrated with buffer E;
     │ elute by a linear phosphate gradient
     │ (2 l) 4.8-340 mM KH_2PO_4 in 0.3 M
     │ KCl, pH 7.0

C-PROTEIN (~100 mg)

CALDESMON

An ubiquituous protein of 86-88 kDa. An asymmetric, highly flexible molecule with 75 nm contour length. Actin binding and inhibitory to actomyosin ATP-ase.

Source:	chicken gizzard smooth muscle
Equipment:	• Waring blender
	• centrifuge
	• water bath
	• Erlenmayer flasks
	• sephacryl S-400 column (1.5 x 170 cm)
	• DEAE-cellulose column (15 ml)
	• electrophoretic gels
Chemicals:	KCl, EGTA, MgCl$_2$, imidazole, DTT, NaN$_3$, NaCl, (NH$_4$)$_2$SO$_4$ (up), ethanol absolute
	protease inhibitors: PMSF, benzamidine
	column materials: sephacryl S-400, DEAE-cellulose
Have ready:	**buffer A:** 0.3 M KCl, 1 mM EGTA, 0.5 mM MgCl$_2$, 50 mM imidazole-HCl, 1 mM DTT, 0.01% NaN$_3$, pH 6.9 [a]
	buffer B: 0.1 M NaCl, 0.1 mM EGTA, 10 mM imidazole-HCl, 1 mM DTT, 0.01% NaN$_3$, pH 7.0 [a]
	buffer C: 30 mM NaCl, 10 mM imidazole-HCl, 1 mM DTT, 0.01% NaN$_3$, pH 7.0

The pH of all buffers is adjusted at 4° C. Since caldesmon is readily oxidized, all buffers should be degassed before the addition of DTT.

[a] These buffers should be made 0.25 mM in PMSF and 1.0 mM in benzamidine immediately before use from a solution in absolute ethanol prepared on the day of use.

Reference:	Lynch, W. and Bretscher, A. (1986) Methods Enzymol. 134, 37-42

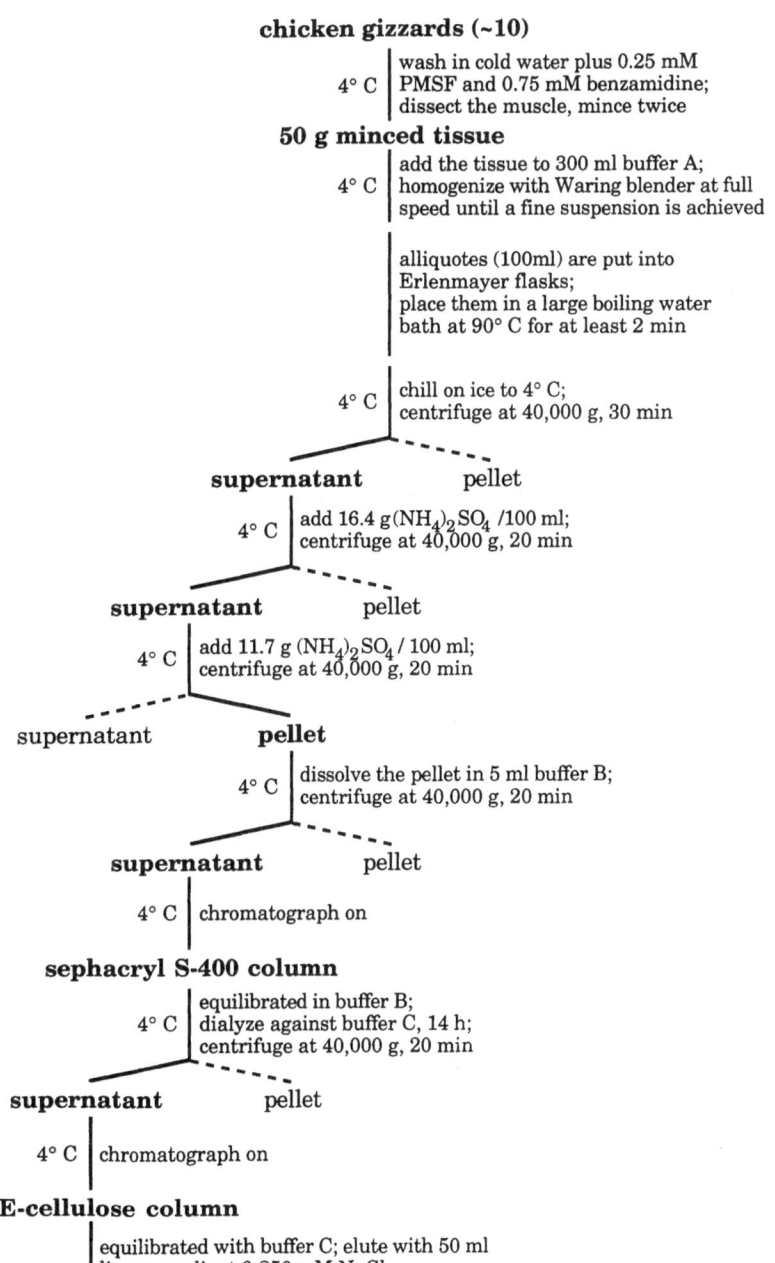

chicken gizzards (~10)

4° C | wash in cold water plus 0.25 mM PMSF and 0.75 mM benzamidine; dissect the muscle, mince twice

50 g minced tissue

4° C | add the tissue to 300 ml buffer A; homogenize with Waring blender at full speed until a fine suspension is achieved

alliquotes (100ml) are put into Erlenmayer flasks; place them in a large boiling water bath at 90° C for at least 2 min

4° C | chill on ice to 4° C; centrifuge at 40,000 g, 30 min

supernatant   pellet

4° C | add 16.4 g $(NH_4)_2SO_4$ /100 ml; centrifuge at 40,000 g, 20 min

supernatant   pellet

4° C | add 11.7 g $(NH_4)_2SO_4$ / 100 ml; centrifuge at 40,000 g, 20 min

supernatant   **pellet**

4° C | dissolve the pellet in 5 ml buffer B; centrifuge at 40,000 g, 20 min

supernatant   pellet

4° C | chromatograph on

sephacryl S-400 column

4° C | equilibrated in buffer B; dialyze against buffer C, 14 h; centrifuge at 40,000 g, 20 min

supernatant   pellet

4° C | chromatograph on

DEAE-cellulose column

equilibrated with buffer C; elute with 50 ml linear gradient 0-250 mM NaCl

CALDESMON

CALPACTIN I and II

Synonymous with annexins, calelectrin and lipocortin. A Ca^{++} dependent phospholipid and actin binding protein of ~35 kDa. Mediator for Ca^{++} signals.

Source: frozen bovine lung or human placenta

Equipment:
- Waring blender
- centrifuge
- sephacryl S-300 column (2.8 x 110 cm)
- two DE-52 columns (10 ml)
- CM-52 column
- two hydroxylapatide columns (3 ml)
- phosphatidylserine acrylamide column (100 ml)
- electrophoretic gels

Chemicals: Trizma-Base, Triton X-100, EGTA, $MgCl_2$, DTT, PMSF, benzamidine, sucrose, $CaCl_2$, NaH_2PO_4, CH_3COONa, imidazole

Have ready: **buffer I:** 40 mM Tris-HCl, 1% Triton X-100, 10 mM EGTA, 2 mM $MgCl_2$, 0.5 mM DTT, 0.2 mM PMSF, 0.5 mM benzamidine, pH 8.8

buffer II: 10 mM imidazole, 2 mM $MgCl_2$, 0.5 mM DTT, 2 mM $CaCl_2$, pH 7.3

buffer III: 20 mM imidazole, 2 mM $MgCl_2$, 0.5 mM DTT, 25 mM EGTA, 200 mM NaCl, pH 7.3

buffer IV: 10 mM imidazole, 25 mM NaCl, 0.5 mM EGTA, 0.5 mM DTT, pH 7.3

buffer V: 20 mM Tris-HCl, 1 mM NaCl, 50 mM EGTA, 1 mM DTT, 2 mM $MgCl_2$, pH 8.0

buffer VI: 20 mM NaH_2PO_4, 0.5 mM DTT, 20 mM EGTA, 2 mM $MgCl_2$, pH 7.3

buffer VII: 10 mM imidazole-HCl, 1 mM EGTA, 1 mM NaN_3, 0.5 mM DTT, pH 7.3

buffer VIII: 10 mM Tris-HCl, 100 mM NaCl, 1 mM EGTA, 0.5 mM DTT, 2 mM $MgCl_2$, pH 8.8

Reference: Glenney, J.R. (1991) Methods Enzymol. 196, 65-69

bovine lung or human placenta (~500 g)

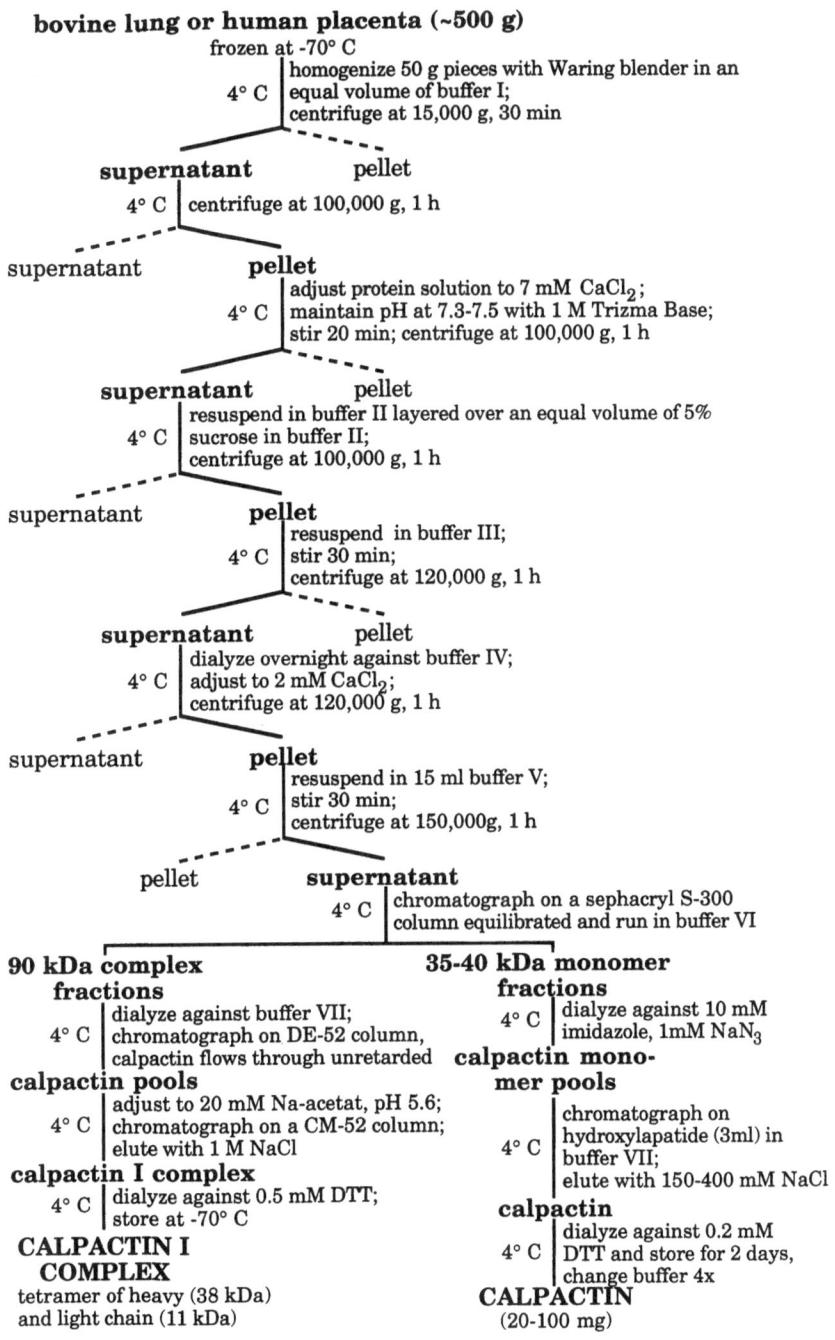

frozen at -70° C

4° C | homogenize 50 g pieces with Waring blender in an equal volume of buffer I; centrifuge at 15,000 g, 30 min

supernatant   pellet

4° C | centrifuge at 100,000 g, 1 h

supernatant   **pellet**

4° C | adjust protein solution to 7 mM $CaCl_2$; maintain pH at 7.3-7.5 with 1 M Trizma Base; stir 20 min; centrifuge at 100,000 g, 1 h

supernatant   pellet

4° C | resuspend in buffer II layered over an equal volume of 5% sucrose in buffer II; centrifuge at 100,000 g, 1 h

supernatant   **pellet**

4° C | resuspend in buffer III; stir 30 min; centrifuge at 120,000 g, 1 h

supernatant   pellet

4° C | dialyze overnight against buffer IV; adjust to 2 mM $CaCl_2$; centrifuge at 120,000 g, 1 h

supernatant   **pellet**

4° C | resuspend in 15 ml buffer V; stir 30 min; centrifuge at 150,000g, 1 h

pellet   **supernatant**

4° C | chromatograph on a sephacryl S-300 column equilibrated and run in buffer VI

90 kDa complex fractions   **35-40 kDa monomer fractions**

4° C | dialyze against buffer VII; chromatograph on DE-52 column, calpactin flows through unretarded

4° C | dialyze against 10 mM imidazole, 1mM NaN_3

calpactin pools   **calpactin mono-mer pools**

4° C | adjust to 20 mM Na-acetat, pH 5.6; chromatograph on a CM-52 column; elute with 1 M NaCl

4° C | chromatograph on hydroxylapatide (3ml) in buffer VII; elute with 150-400 mM NaCl

calpactin I complex   **calpactin**

4° C | dialyze against 0.5 mM DTT; store at -70° C

4° C | dialyze against 0.2 mM DTT and store for 2 days, change buffer 4x

CALPACTIN I COMPLEX
tetramer of heavy (38 kDa) and light chain (11 kDa)

CALPACTIN
(20-100 mg)

55

CALPONIN

A 34 kDa actin and calmodulin binding protein. Possibly regulating smooth muscle contraction, monomeric with various isoforms.

Source: avian gizzard and porcine stomach

Equipment:
- mincer
- microwave oven
- mason jar
- Sorvall Omnimixer
- centrifuge
- glass wool
- S-sepharose fast flow column
- A-superose 2 FPLC column
- electrophoretic gels

Chemicals: MES, NaCl, EDTA, EGTA, KCl, $(NH_4)_2SO_4$, imidazole, $MgCl_2$, DTE

Have ready: **S-buffer:** 25 mM MES, 25 mM NaCl, 0.5 mM EDTA, 0.5 mM EGTA, pH 5.4

Reference: Vancompernolle, K., Gimona, M., Herzog, M., van Damme, J., Vandekerckhove, J., and Small, J.V. (1990) FEBS letters 274, 146-150
Gen-Bank / EMBL: M63559, M63560

56

**avian gizzard or porcine stomach
fresh muscle tissue (500 g)**

clean and mince the tissue;
heat the tissue in a microwave oven for 5 min;
transfer into a mason jar;
poure ice cold S-buffer+ 300 mM KCl onto the heated muscle;
homogenize in a Sorvall Omnimixer at top speed for 1 min

4° C — extract 30 min;
centrifuge at 11,000 rpm, 10 min (Sorvall RC-5 with GSA rotor)

supernatant   pellet

4° C — filter through glass wool;
add 30% $(NH_4)_2SO_4$; stir 10 min;
centrifuge at 11,000 rpm, 10 min (Sorvall RC-5 with GSA rotor)

supernatant   **pellet**

4° C — dissolve the pellet in S-buffer;
dialyze overnight against S-buffer;
centrifuge at 18,000 rpm, 20 min (Sorvall RC-5 with GSA rotor)

supernatant   pellet

4° C — apply to

S-sepharose fast flow column

4° C — equilibrated in S-buffer;
chromatograph on

A-superose 2 FPLC column

CALPONIN

CAP-100

A 100 kDa barbed end F-actin capping protein without nucleating activity. A Ca^{++}-independent but PIP-2 inhibited G-actin and F-actin binding protein acting at substoichiometric concentrations (<1:100).

Source: Dictyostelium discoideum

Equipment:
- Parr bomb
- DEAE-cellulose (DE-52) column (5 x 40 cm)
- phosphocellulose (P-11) column (2.5 x 15 cm)
- hydroxylapatide column (2.5 x 5 cm)
- sepharose-CL-6B column (2.5 x 100 cm)
- Mono-Q column (FPLC, Pharmacia)

Chemicals: sucrose, EDTA, EGTA, Tris-HCl, ATP, DTT, NaN$_3$, MES, K/PO$_4$, (NH$_4$)$_2$SO$_4$, imidazole, NaCl
protease inhibitors: benzamidine, PMSF

Have ready: **Soerensen phosphate buffer,** pH 6.0

homogenization buffer: 30 mM Tris-HCl, 4 mM EGTA, 2 mM EDTA, 2 mM DTT, 0.2 mM ATP, 5 mM benzamidine, 0.5 mM PMSF, 30% sucrose, pH 8.0

DEAE-column buffer: 10 mM Tris-HCl, 1 mM EGTA, 1 mM DTT, 0.02% NaN$_3$, 1 mM benzamidine, 0.5 mM PMSF, 0.1 mM ATP, pH 8.0

P-11 column buffer: 10 mM MES, 1 mM EGTA, 1 mM DTT, 0.1 mM ATP, 0.02% NaN$_3$, 1 mM benzamidine, 0.5 mM PMSF, pH 6.5

hydroxylapatide (HT) column buffer: 10 mM potassium phosphate, 1 mM EGTA, 0.2 mM DTT, 0.1 mM ATP, 0.02% NaN$_3$, 1 mM benzamidine, 0.5 mM PMSF, pH 6.5

sepharose-CL-6B column buffer: 10 mM imidazole, 1 mM EGTA, 1 mM DTT, 0.02% NaN$_3$, 0.1 mM ATP, 1 mM benzamidine, 0.5 mM PMSF, 0.2 M NaCl, pH 7.6

Mono-Q column buffer: 30 mM Tris-HCl, 0.1 mM EGTA, 0.15 mM DTT, 0.02% NaN$_3$, pH 7.8

Reference: Hofmann, A. et al. (1992) Cell Mot. and Cytoskel. 23, 133-144

Dictyostelium discoideum (~500 g)

0-4° C | wash axenically grown cells (30 l fermenters)
twice in Soerensen phosphate buffer, pH 6.0;
homogenize in homogenization buffer;
open cells with Parr bomb (4 MPa per 15 min);
centrifuge at 10,000 g, 30 min

supernatant   pellet

0-4° C | reextract and centrifuge at 100,000 g, 3 h

supernatant   pellet

0-4° C | adjust pH to 8.0;
load onto

DEAE-cellulose column

0-4° C | preequilibrated with DEAE column buffer;
elute with linear gradient (2 x 750 ml) 0-300 mM NaCl
in column buffer; pool fractions at 2-5 mS;
add solid MES to 10 mM; lower pH to 6.5 and directly apply to

phosphocellulose (P-11) column

0-4° C | preequilibrated with P-11 column buffer;
elute with linear gradient (2 x 200 ml) 0-400 mM
NaCl in column buffer; pool fractions around ~11 mS;
dialyze against HT column buffer;
load onto

hydroxylapatide column

0-4° C | preequilibrated with HT-column buffer;
elute with linear gradient (2 x 120 ml) 0-300 mM potassium
phosphate; pool fractions around 10-13 mS; reduce volume by
65% $(NH_4)_2SO_4$ precipitation and apply to

sepharose CL-6B column

0-4° C | preequilibrated with sepharose CL-6B column buffer;
elute at 20 ml/h; apply eluate to

Mono-Q column

0-4° C | elute with salt gradient 0-0.5 M NaCl in column buffer;
collect fractions at ~200 mM NaCl

CAP-100 (~300-800 µg)

CAPPING PROTEINS

Capping proteins are heterodimers of 36 kDa and 32 kDa. They bind Ca^{++}- independent to the fast growing end of actin filaments and nucleate polymerization.

Source: chicken skeletal muscle

Equipment:
- cheesecloth
- DEAE-cellulose column (2.5 x 50 cm)
- hydroxylapatide column (1 x 15 cm)
- sephacryl S-200 column (1.6 x 90 cm)
- electrophoretic gels
- centrifuge
- centrifuge with a vertical rotor (sucrose gradient)
- Amicon cell

Chemicals: $(NH_4)_2SO_4$, DMS, ATP, Trizma Base, $CaCl_2$, NaN_3, 2-mercaptoethanol, $Na_2S_2O_3$, KCl, DTT, KH_2PO_4, sucrose
protease inhibitors: DFP, PMSF, pepstatin, leupeptin, aprotinin,

Have ready: **buffer A:** 0.2 mM ATP, 0.2 mM $CaCl_2$, 0.5 mM 2-mercaptoethanol, 0.01% (w/v) NaN_3, 5 mM Tris-HCl, pH 8.0 (5 liters)

buffer B: 0.6 M KI, 20 mM $Na_2S_2O_3$, 5 mM 2-mercaptoethanol, 0.01% (w/v) NaN_3, 10 mM Tris-HCl, pH 7.2 (3 liters)

buffer C: 50 mM KCl, 1 mM 2-mercaptoethanol, 0.01% (w/v) NaN_3, 10 mM Tris-HCl, pH 8.0
same buffer with 500 mM KCl (2 liters)

buffer D: 1 M KCl, 0.1 mM DTT, 0.01% (w/v) NaN_3, 1 mM KH_2PO_4, pH 7.0 (3 liters)
same buffer with 75 mM KH_2PO_4 (100 ml)

buffer E: 20% sucrose, 100 mM KCl, 0.01% (w/v) NaN_3, 10 mM KH_2PO_4, pH 8.0 (1 liter)
same buffer with 5% sucrose (100 ml)

1 M KCl (2 liters)

protease inhibitors: 0.5 mM PMSF, 0.4 mM DFP, 5 mg/ml aprotinin, 5 mg/ml leupeptin, 1 mg/ml pepstatin

Reference: Isenberg, G., Aebi, U. and Pollard, T.D. (1980) Nature 288, 455-459
Casella, J.F. and Cooper, J.A. (1991) Methods Enzymol. 196, 140-154
Gen-Bank: M36882 (chicken), M31720(Saccharomyces c.)

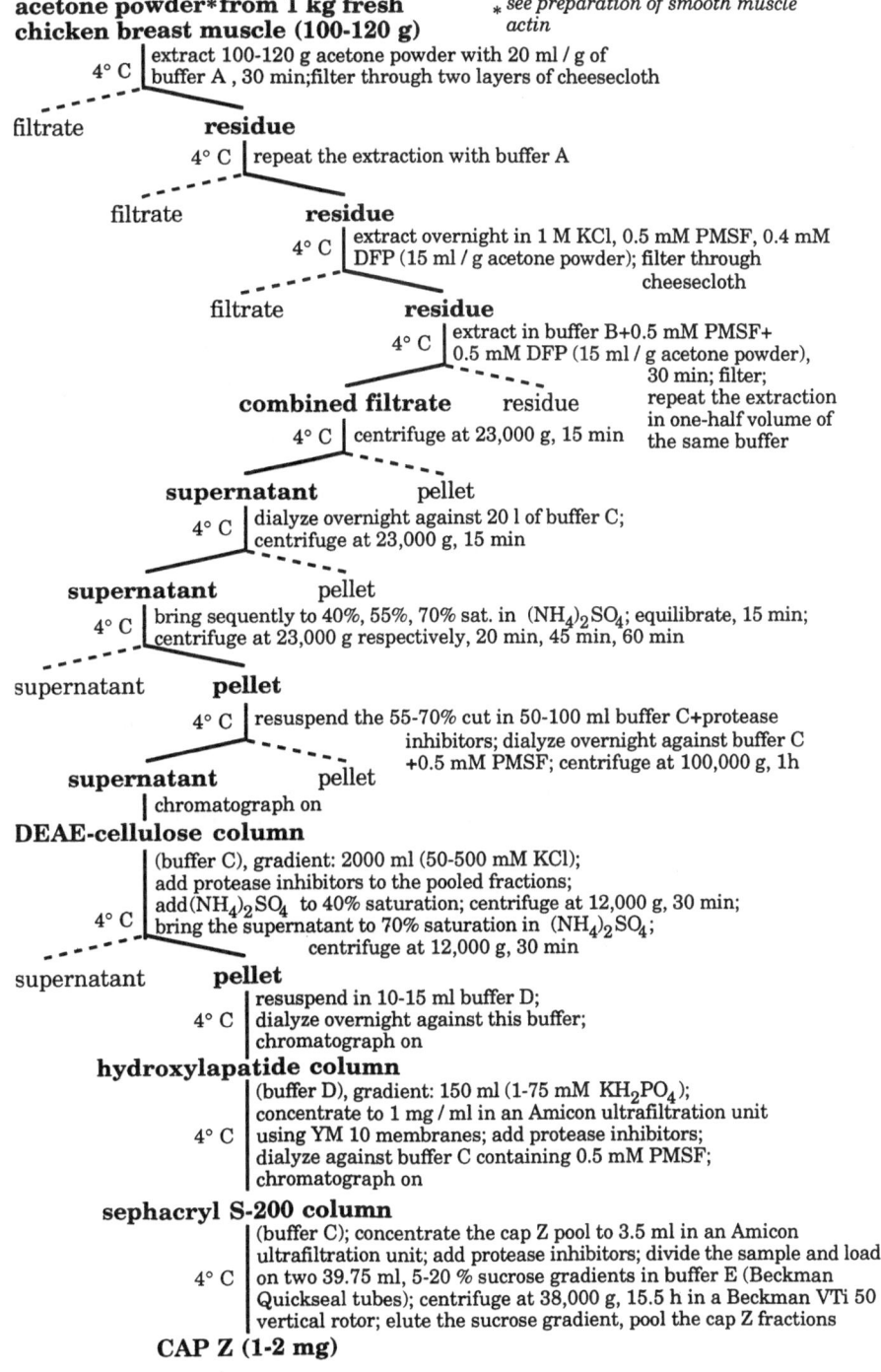

acetone powder*from 1 kg fresh chicken breast muscle (100-120 g)

* *see preparation of smooth muscle actin*

4° C | extract 100-120 g acetone powder with 20 ml / g of buffer A , 30 min;filter through two layers of cheesecloth

filtrate — **residue**

4° C | repeat the extraction with buffer A

filtrate — **residue**

4° C | extract overnight in 1 M KCl, 0.5 mM PMSF, 0.4 mM DFP (15 ml / g acetone powder); filter through cheesecloth

filtrate — **residue**

4° C | extract in buffer B+0.5 mM PMSF+ 0.5 mM DFP (15 ml / g acetone powder), 30 min; filter; repeat the extraction in one-half volume of the same buffer

combined filtrate — residue

4° C | centrifuge at 23,000 g, 15 min

supernatant — pellet

4° C | dialyze overnight against 20 l of buffer C; centrifuge at 23,000 g, 15 min

supernatant — pellet

4° C | bring sequently to 40%, 55%, 70% sat. in $(NH_4)_2SO_4$; equilibrate, 15 min; centrifuge at 23,000 g respectively, 20 min, 45 min, 60 min

supernatant — **pellet**

4° C | resuspend the 55-70% cut in 50-100 ml buffer C+protease inhibitors; dialyze overnight against buffer C +0.5 mM PMSF; centrifuge at 100,000 g, 1h

supernatant — pellet

| chromatograph on

DEAE-cellulose column

4° C | (buffer C), gradient: 2000 ml (50-500 mM KCl); add protease inhibitors to the pooled fractions; add$(NH_4)_2SO_4$ to 40% saturation; centrifuge at 12,000 g, 30 min; bring the supernatant to 70% saturation in $(NH_4)_2SO_4$; centrifuge at 12,000 g, 30 min

supernatant — **pellet**

4° C | resuspend in 10-15 ml buffer D; dialyze overnight against this buffer; chromatograph on

hydroxylapatide column

4° C | (buffer D), gradient: 150 ml (1-75 mM KH_2PO_4); concentrate to 1 mg / ml in an Amicon ultrafiltration unit using YM 10 membranes; add protease inhibitors; dialyze against buffer C containing 0.5 mM PMSF; chromatograph on

sephacryl S-200 column

4° C | (buffer C); concentrate the cap Z pool to 3.5 ml in an Amicon ultrafiltration unit; add protease inhibitors; divide the sample and load on two 39.75 ml, 5-20 % sucrose gradients in buffer E (Beckman Quickseal tubes); centrifuge at 38,000 g, 15.5 h in a Beckman VTi 50 vertical rotor; elute the sucrose gradient, pool the cap Z fractions

CAP Z (1-2 mg)

COFILIN

A 21 kDa protein of 166 amino acids which binds to G- and F-actin in a 1:1 molar ratio and deploymerizes actin filaments in a pH and lipid binding dependent manner. Present in many avian and mammalian cells.

Source:	porcine brain
Equipment:	• blender
	• centrifuge
	• DNA-se I Affi gel-10 column (2.2 x 17 cm)
	• hydroxylapatide column (2 x 22 cm)
	• Whatman P-11 cellulose column (1.4 x 7 cm)
	• sephadex G-75 column (1.1 x 38 cm)
Chemicals:	Tris-HCl, EGTA, ATP, DTT, PMSF, DNA-se I-Affi gel 10, HEPES-KOH, $MgCl_2$, $CaCl_2$, K/PO_4, PIPES-KOH, NaCl
Have ready:	**extraction buffer:** 10 ml Tris-HCl, 1 mM EGTA, 1 mM DTT, 1 mM PMSF, 1 mM ATP, pH 7.8
	dialyzing buffer 1: 2 mM HEPES-KOH, 0.1 mM ATP, 0.1 mM DTT, pH 7.2
	DNA-se I column buffer: 2 mM HEPES-KOH, 1 mM $MgCl_2$, 1 mM $CaCl_2$, pH 7.2
	hydroxylapatide column buffer: 10 mM K/PO_4, 1 mM $MgCl_2$, 0.1 M NaCl, 0.1 mM DTT, pH 6.8
	dialyzing buffer 2: 10 mM PIPES-KOH, 0.1 mM DTT, pH 7.0
	sephadex G-75 column buffer: 10 mM PIPES-KOH, 0.1 mM DTT, 100 mM KCl, pH 7.3
Reference:	Maekawa et al. (1984) J. Biochem. 95, 377-385 Gen-Bank: J03917 (pig), D00472 (mouse), J02915 (chicken), D00682 (human)

6 porcine brains (~550 g)

| homogenize in 300 ml of extraction buffer;
| centrifuge at 20,000 g, 60 min

supernatant   pellet

| dialyze overnight against dialyzing buffer 1 (2 x);
| centrifuge at 100,000 g, 90 min

supernatant   pellet
| add 1 mM $MgCl_2$, 1 mM $CaCl_2$;
| mix with Affi-gel 10 equilibrated with DNA-se I
| column buffer and stir at 0°C for 1 h; apply to

DNA-se I column

| elute with 600 mM NaCl in column buffer;
| apply 21 kDa protein fractions to

hydroxylapatide column

| elute with linear gradient 10-150 mM K/PO_4 + 1 mM $MgCl_2$
| and 0.1 mM DTT, pH 6.8;
| collect fractions around 70 mM phosphate;
| dialyze against dialyzing buffer 2 and apply to

P-11 cellulose column

| wash, then elute with linear gradient 0-200 mM NaCl in dialysis buffer 2;
| collect fractions around 50 mM;
| concentrate with Amicon PM 10 membrane and apply to

sephadex G-75 column

| elute with sephadex G-75 column buffer

COFILIN (~1 mg)

COMITIN

A 24 kDa membrane and F-actin binding protein accumulated in perinuclear Golgi-membranes in a variety of cells.

Source:	Dictyostelium discoideum
Equipment:	• microwave radiation or Parr bomb • centrifuge • airfuge • sonicator • vacuum dialysis system • BioGel A 1.5 m column (1.5 x 45 cm)
Chemicals:	Tris-HCl, sucrose, $Na_4P_2O_7$, NaN_3, EDTA, DTT, ethanol, imidazole, ATP, DOC (sodium deoxycholate) **protease inhibitors:** PMSF, N-α-p-tosyl-L-lysyl-chloromethylketone, 1.10-phenanthroline, N-carbobenzoxy-L-phenylalanine
Have ready:	**lysis buffer:** 10 mM Tris-HCl, 30% sucrose, 40 mM sodium pyrophosphate, 2 mM EDTA, 0.2 mM DTT, 0.02% NaN_3, 2 mM N-α-p-tosyl-L-lysyl-chloromethylketone, 5 mM 1.10-phenanthroline, 0.1 mg/ml PMSF, 2 mM N-carbobenzoxy-L-phenylalanine, 0.5% ethanol, pH 7.6
	G-buffer: 10 mM imidazole, 0.2 mM ATP, 0.2 mM DTT, pH 7.4
	TED-buffer: 10 mM Tris-HCl, 1 mM EDTA, 0.2 mM DTT, pH 8.1
Reference:	Stratford, C.A. and Brown, S.S. (1985) J. Cell Biol. 100, 727-735 Weiner et al. (1993) J. Cell Biol. 123, 23-34

Dictyostelium discoideum
(10^7 cells/ml, 600 g wet weight)

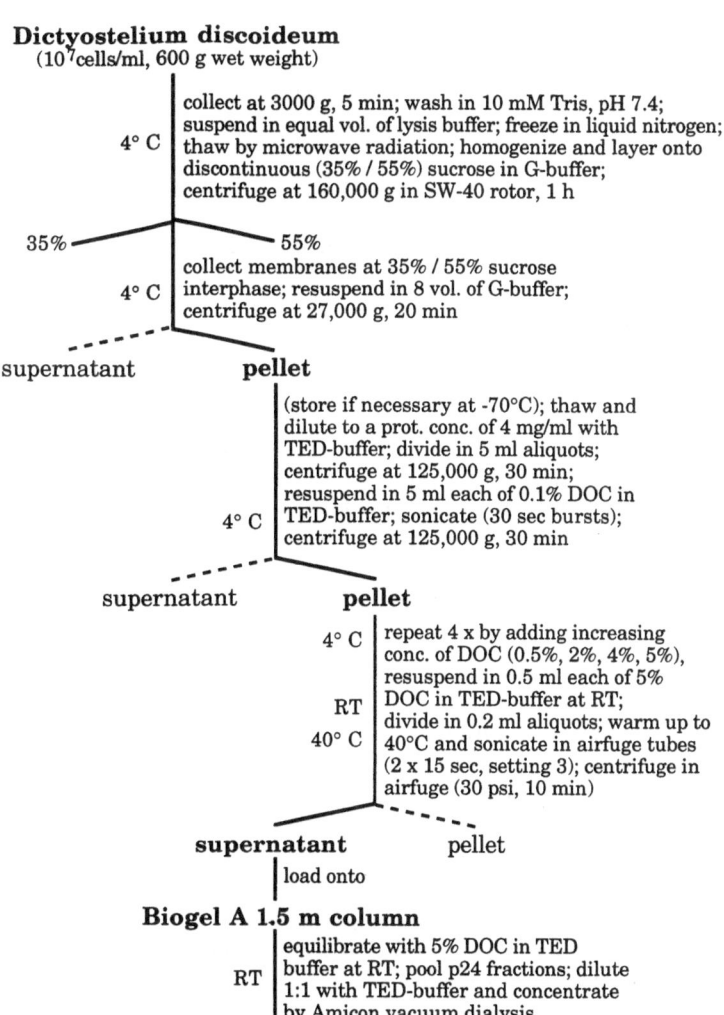

4° C collect at 3000 g, 5 min; wash in 10 mM Tris, pH 7.4; suspend in equal vol. of lysis buffer; freeze in liquid nitrogen; thaw by microwave radiation; homogenize and layer onto discontinuous (35% / 55%) sucrose in G-buffer; centrifuge at 160,000 g in SW-40 rotor, 1 h

35% — 55%

4° C collect membranes at 35% / 55% sucrose interphase; resuspend in 8 vol. of G-buffer; centrifuge at 27,000 g, 20 min

supernatant   **pellet**

4° C (store if necessary at -70°C); thaw and dilute to a prot. conc. of 4 mg/ml with TED-buffer; divide in 5 ml aliquots; centrifuge at 125,000 g, 30 min; resuspend in 5 ml each of 0.1% DOC in TED-buffer; sonicate (30 sec bursts); centrifuge at 125,000 g, 30 min

supernatant   **pellet**

4° C

RT

40° C repeat 4 x by adding increasing conc. of DOC (0.5%, 2%, 4%, 5%), resuspend in 0.5 ml each of 5% DOC in TED-buffer at RT; divide in 0.2 ml aliquots; warm up to 40°C and sonicate in airfuge tubes (2 x 15 sec, setting 3); centrifuge in airfuge (30 psi, 10 min)

supernatant   pellet
load onto

Biogel A 1.5 m column

RT equilibrate with 5% DOC in TED buffer at RT; pool p24 fractions; dilute 1:1 with TED-buffer and concentrate by Amicon vacuum dialysis

COMITIN (p24)

DEMATIN (PROTEIN 4.9)

A 145.000 MW trimer consisting of 48 and 52 kDa subunits comprising the membrane skeleton of erythrocytes. In vitro an effective actin bundling protein which is inhibited by phosphorylation through protein kinase A.

Source: erythrocytes

Equipment:
- DEAE-sephacel column (2.5 x 42 cm)
- hydroxylapatide Ultrogel column (1 x 12 cm)
- centrifuge
- SDS-PAGE

Chemicals: Triton X-100, EGTA, EDTA, DTT, Tris-HCl, NaN_3, Na/PO$_4$, NaCl, K/PO$_4$, di-isopropyl-fluorophosphate, $(NH_4)_2SO_4$

protease inhibitors: PMSF, trasylol, benzamidine, antipain, chymostatin, pepstatin A, leupeptin

Have ready: **lysis buffer:** 5 mM Na/PO$_4$, 1 mM EDTA, 0.5 mM PMSF, pH 8.0

washing buffer: 155 mM NaCl, 2 mM di-isopropyl fluorophosphate

extraction buffer: 0.5% Triton X-100, 7.5 mM Na/PO$_4$, 1 mM EGTA, 0.5 mM PMSF, 3 units/ml trasylol, 10 µg/ml benzamidine, 2 µg/ml antipain, 1 µg/ml each of chymostatin, pepstatin A, leupeptin, pH 8.0

dissociation buffer: 3 mM Tris-HCl, 0.5 mM EDTA, 2 mM DTT, 0.5 mM PMSF plus protease inhibitors (see above), pH 8.5

DEAE-Sephacel column buffer: 20 mM NaCl, 1 mM EGTA, 0.5 mM DTT, 0.02% NaN_3, 20 mM Tris-HCl, pH 8.3

hydroxylapatide Ultrogel column buffer: 20 mM NaCl, 0.2 mM PMSF, 0.5 mM DTT, 10 mM K/PO$_4$, pH 7.2

Reference: Athar Husain-Chishti et al. (1989) J. Biol. Chem. 264, 8985-8991

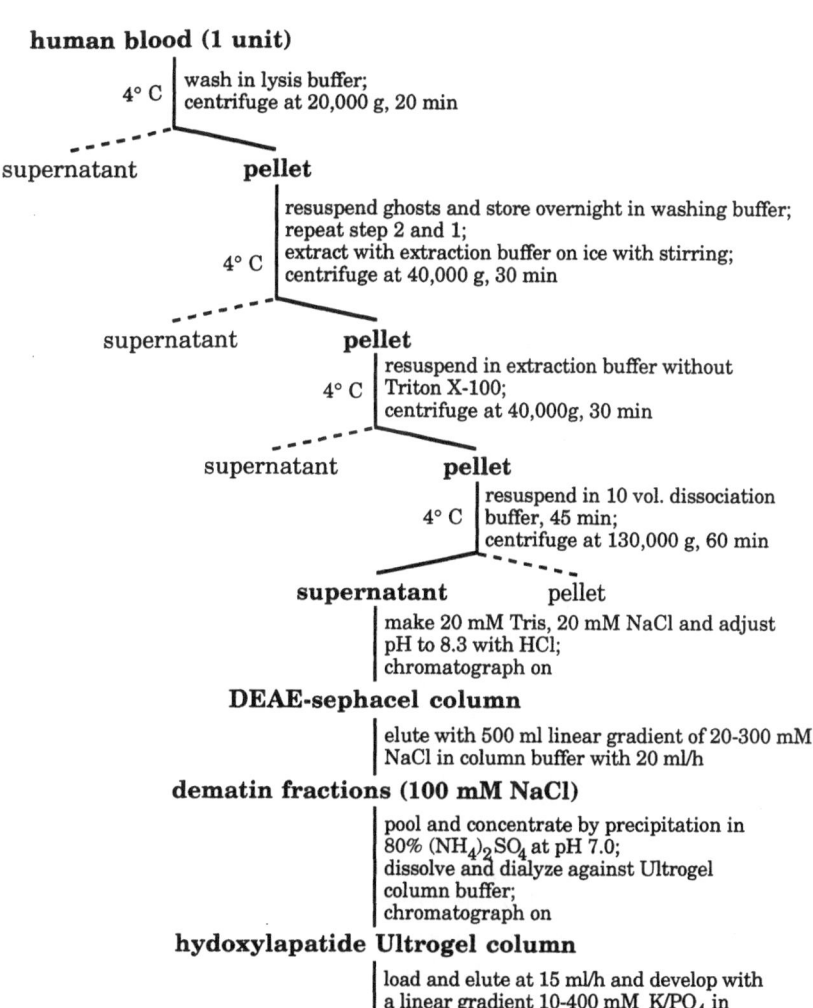

human blood (1 unit)

4° C | wash in lysis buffer;
centrifuge at 20,000 g, 20 min

supernatant   **pellet**

4° C | resuspend ghosts and store overnight in washing buffer;
repeat step 2 and 1;
extract with extraction buffer on ice with stirring;
centrifuge at 40,000 g, 30 min

supernatant   **pellet**

4° C | resuspend in extraction buffer without
Triton X-100;
centrifuge at 40,000g, 30 min

supernatant   **pellet**

4° C | resuspend in 10 vol. dissociation
buffer, 45 min;
centrifuge at 130,000 g, 60 min

supernatant   pellet

make 20 mM Tris, 20 mM NaCl and adjust
pH to 8.3 with HCl;
chromatograph on

DEAE-sephacel column

elute with 500 ml linear gradient of 20-300 mM
NaCl in column buffer with 20 ml/h

dematin fractions (100 mM NaCl)

pool and concentrate by precipitation in
80% $(NH_4)_2SO_4$ at pH 7.0;
dissolve and dialyze against Ultrogel
column buffer;
chromatograph on

hydoxylapatide Ultrogel column

load and elute at 15 ml/h and develop with
a linear gradient 10-400 mM K/PO_4 in
column buffer

DEMATIN

DEPACTIN

A 17 kDa actin depolymerizing protein equivalent to actophorin from
Acanthamoeba which binds to G-actin in a 1:1 molar complex.

Source: starfish eggs

Equipment:
- centrifuge
- sephadex G-150 column (5 x 85 cm)
- DEAE cellulose column
- hydroxylapatide column
- polyacrylamide gels

Chemicals: glycerol, $NaHCO_3$, ATP, NaCl, glucose, $MgCl_2$, KCl, K/PO_4, DTT, MOPS·NaOH (morpholino-propane sulfonic acid), TAME (p-tosyl-L-arginine methyl ester-HCl), PIPES (piperazine-N-N′-bis (2-ethane sulfonic acid)), Tris-HCl, EGTA, Aquacite IA

Have ready: **washing buffer:** 0.5 M glycerol, 0.2 M NaCl, 10 mM $NaHCO_3$

extraction buffer: 0.7 M glucose, 0.1 M KCl, 2 mM $MgCl_2$, 5 mM EGTA, 10 mM MOPS·NaOH, 10 mM TAME, 1 mM ATP, 1 mM DTT, pH 6.85

gel filtration buffer: 0.1 M KCl, 2 mM $MgCl_2$, 1 mM EGTA, 10 mM MOPS·NaOH, 1 mM TAME, 0.2 mM ATP, 0.2 mM DTT, pH 6.85

DEAE column buffer: 10 mM Tris-HCl, 0.5 mM EGTA, 0.5 mM TAME, 0.5 mM DTT, pH 8.2

hydroxylapatide column buffer: 10 mM K/PO_4, 0.5 mM TAME, 0.5 mM DTT, pH 6.8

Reference: Mabuchi, I. (1981) J. Biochem. 89, 1341-1344, 1981

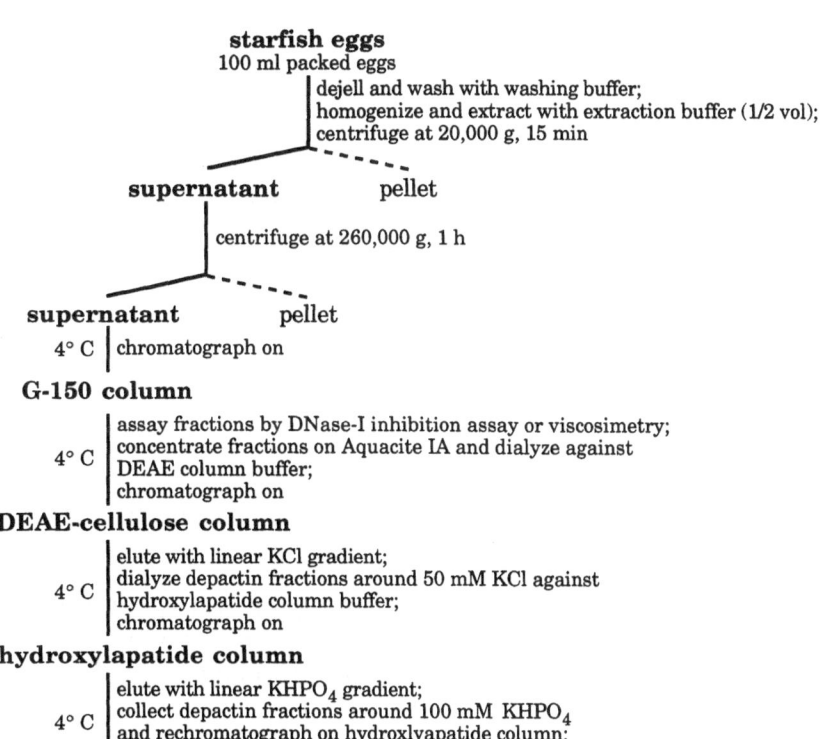

starfish eggs
100 ml packed eggs

dejell and wash with washing buffer;
homogenize and extract with extraction buffer (1/2 vol);
centrifuge at 20,000 g, 15 min

supernatant   pellet

centrifuge at 260,000 g, 1 h

supernatant   pellet

4° C │ chromatograph on

G-150 column

4° C │ assay fractions by DNase-I inhibition assay or viscosimetry;
concentrate fractions on Aquacite IA and dialyze against
DEAE column buffer;
chromatograph on

DEAE-cellulose column

4° C │ elute with linear KCl gradient;
dialyze depactin fractions around 50 mM KCl against
hydroxylapatide column buffer;
chromatograph on

hydroxylapatide column

4° C │ elute with linear KHPO$_4$ gradient;
collect depactin fractions around 100 mM KHPO$_4$
and rechromatograph on hydroxlyapatide column;
rechromatograph on DEAE-cellulose column

DEPACTIN (~1 mg)

DESTRIN

A 19 kDa monomeric actin depolymerizing protein equivalent to ADF, depactin and actophorin. 165 amino acids with sequence similarity to cofilin. A pH independent but PIP-inhibited severing protein.

Source:	porcine brain
Equipment:	• mixer • DNA-se I-Affi gel-10 column (2.2 x 17 cm) • hydroxylapatide column (2 x 22 cm) • Amicon concentration device • sephadex G-75 column (1.1 x 38 cm)
Chemicals:	Tris-HCl, DTT, EGTA, ATP, HEPES, MgCl$_2$, CaCl$_2$, DNA-se I, Affi gel-10, urea, K/PO$_4$, NaCl, KCl, PIPES-KOH **protease inhibitors:** PMSF
Have ready:	**homogenization solution:** 10 mM Tris-HCl, 1 mM EGTA, 1 mM DTT, 1 mM PMSF, 1 mM ATP, pH 7.8 **dialyzing buffer:** 2 mM HEPES-KOH, 0.1 mM ATP, 0.1 mM DTT, pH 7.2 **washing buffer II:** 10 mM HEPES-KOH, 1 mM MgCl$_2$, 1 mM CaCl$_2$, pH 7.2 **elution buffer I:** 10 mM HEPES, 1 mM MgCl$_2$, 2 M urea, 0.1 mM ATP, pH 7.2 **elution buffer II:** elution buffer I with 6 M urea **hydroxylapatide column buffer:** 10 mM K/PO$_4$, 1 mM MgCl$_2$, 0.1 M NaCl, 0.1 mM DTT, pH 6.8 **gel filtration buffer:** 10 mM PIPES-KOH, 100 mM KCl, 0.1 mM DTT, pH 7.3
Reference:	Maekawa, S. et al. (1984) J. Biochem. 95, 377-385 Nishida, E. et al. (1984) J. Biochem. 95, 387-398

6 porcine brains (~550 g)

$\quad$ homogenize with mixer in
4° C $\quad$ homogenization solution (300 ml);
$\quad$ centrifuge at 20,000 g, 60 min

supernatant $\qquad$ pellet

dialyze overnight against dialyzing buffer;
clarify at 100,000 g, 90 min;
add 1 mM $MgCl_2$, 1 mM $CaCl_2$;
mix with 100 ml DNA-se I-Affi gel-10
equilibrated in the last buffer; stir for 1 h;
pour

DNA-se I-Affi gel column

wash with washing buffer II;
elute stepwise with
- 600 mM NaCl in washing buffer II,
- washing buffer II + 2 mM EGTA, no $CaCl_2$!,
- elution buffer I,
- elution buffer II;
collect 12-15 ml fractions;
pool salt-step eluate, dialyze against hydroxylapatide
column buffer and apply to

hydroxylapatide column

preequilibrate with hydroxylapatide column buffer;
elute with 10-150 mM K/PO_4 in column buffer;
pool protein peak eluting at 10 mM phosphate;
concentrate with Amicon PM-10 membrane and apply to

sephadex G-75 column

equilibrate and run in gel filtration buffer

DESTRIN (~0.3 mg)

DYSTROPHIN

A 427 kDa membrane protein from muscle. A gene defect of
dystrophin causes Duchenne muscular dystrophy (DMD). Precise
function still unknown. Sequence similarities of actin binding domains
with α-actinin and spectrin.

Source:	rabbit skeletal muscle
Equipment:	• meat grinder
	• Waring blender
	• cheese cloth
	• SW 28 swing out rotor, Ti 45 rotor
	• centrifuge
	• XA 7-sepharose column
	• WGA-sepharose column (15 ml)
	• DEAE-cellulose column
	• 3-12% gradient SDS-gels
Chemicals:	Tris-HCl, $Na_4P_2O_7$, Na_2HPO_4, $MgCl_2$, EDTA, sucrose, Tris-maleate, digitonin, NAG (N-acetyl-D-glucosamine), KCl
	protease inhibitors: aprotinin, iodacetamine, PMSF, benzamidine
Have ready:	**buffer A:** 20 mM $Na_4P_2O_7$, 20 mM Na_2HPO_4, 1 mM $MgCl_2$, 0.303 M sucrose, 0.5 mM EDTA, pH 7.0 incl. 0.83 nM aprotinin, 1 mM iodacetamine, 0.23 mM PMSF
	buffer B: 0.303 M sucrose, 20 mM Tris-maleate, pH 7.0
	sucrose gradient: 0.878 M sucrose, 0.6 M KCl, 20 mM Tris-maleate, pH 7.0
	WGA-sepharose column buffer: 50 mM Tris-HCl, pH 7.4, 0.75 mM benzamidine, 0.1 mM PMSF
Reference:	Campbell, K.P. and Kahl, S.D. (1989) Nature 338, 259-262
	Gen-Bank: Gen EMBL M18533 (human), X13369 (chicken)

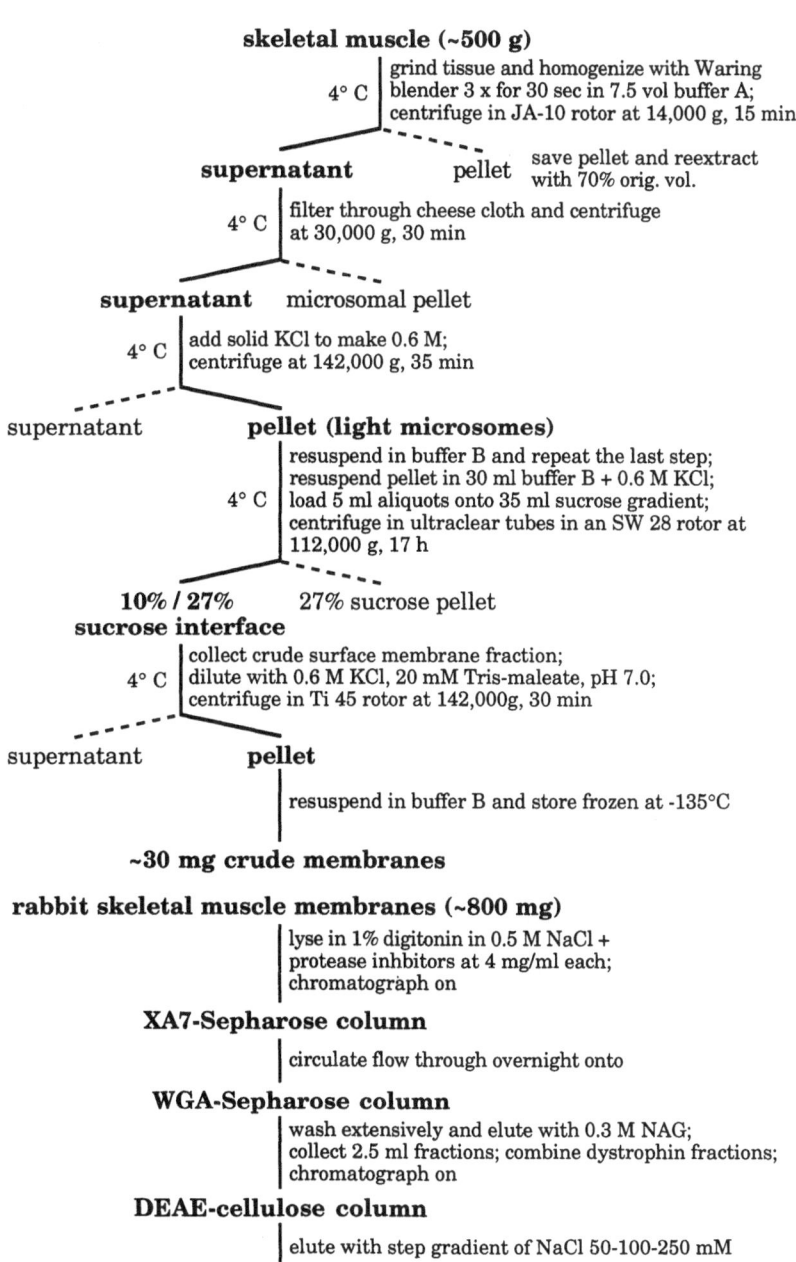

skeletal muscle (~500 g)

4° C | grind tissue and homogenize with Waring blender 3 x for 30 sec in 7.5 vol buffer A; centrifuge in JA-10 rotor at 14,000 g, 15 min

supernatant   pellet — save pellet and reextract with 70% orig. vol.

4° C | filter through cheese cloth and centrifuge at 30,000 g, 30 min

supernatant   microsomal pellet

4° C | add solid KCl to make 0.6 M; centrifuge at 142,000 g, 35 min

supernatant   **pellet (light microsomes)**

4° C | resuspend in buffer B and repeat the last step; resuspend pellet in 30 ml buffer B + 0.6 M KCl; load 5 ml aliquots onto 35 ml sucrose gradient; centrifuge in ultraclear tubes in an SW 28 rotor at 112,000 g, 17 h

10% / 27% sucrose interface   27% sucrose pellet

4° C | collect crude surface membrane fraction; dilute with 0.6 M KCl, 20 mM Tris-maleate, pH 7.0; centrifuge in Ti 45 rotor at 142,000g, 30 min

supernatant   **pellet**

| resuspend in buffer B and store frozen at -135°C

~30 mg crude membranes

rabbit skeletal muscle membranes (~800 mg)

| lyse in 1% digitonin in 0.5 M NaCl + protease inhbitors at 4 mg/ml each; chromatograph on

XA7-Sepharose column

| circulate flow through overnight onto

WGA-Sepharose column

| wash extensively and elute with 0.3 M NAG; collect 2.5 ml fractions; combine dystrophin fractions; chromatograph on

DEAE-cellulose column

| elute with step gradient of NaCl 50-100-250 mM

DYSTROPHIN

EZRIN

A membrane associated monomeric protein of 80 kDa possibly linking
F-actin to membranes. 585 amino acids with 34% sequence identity to
protein 4.1

Source:	brush borders from chicken intestines
Equipment:	• sephadex G-50 column (100 ml) • hydroxylapatide column (10 ml) • DEAE-cellulose column (0.5 ml) • electrophoretic gels, 5-20% gradient gels • centrifuge
Chemicals:	KCl, EGTA, $MgCl_2$, imidazole, guanidinium, Na-acetat, $CaCl_2$, NaCl, KH_2PO_4, DTT, $(NH_4)_2SO_4$ (up) **protease inhibitors:** PMSF, benzamidin **column materials:** sephadex G-50, DEAE-cellulose, hydroxylapatide
Have ready:	**buffer H:** 75 mM KCl, 1 mM EGTA, 0.1 mM $MgCl_2$, 0.25 mM PMSF, 0.5 mM benzamidine, 10 mM imidazole-HCl, pH 7.3
	buffer M: 3 M guanidinium-HCl, 0.5 M Na-acetat, 1 mM $CaCl_2$, 5 mM $MgCl_2$, 0.5 M NaCl, 0.25 mM PMSF, 0.5 mM benzamidine, pH 6.5
	buffer N: 50 mM KH_2PO_4, 0.1 mM $CaCl_2$, 1 mM DTT, pH 7.0
	buffer O: 20 mM NaCl, 10 mM imidazole-HCl, 1 mM DTT, pH 6.7
Reference:	Bretscher, A. (1986) Methods Enzymol. 134, 34-37 Gen-Bank: X51521

chicken brush borders (10 chickens)
resuspend in 100 ml buffer H

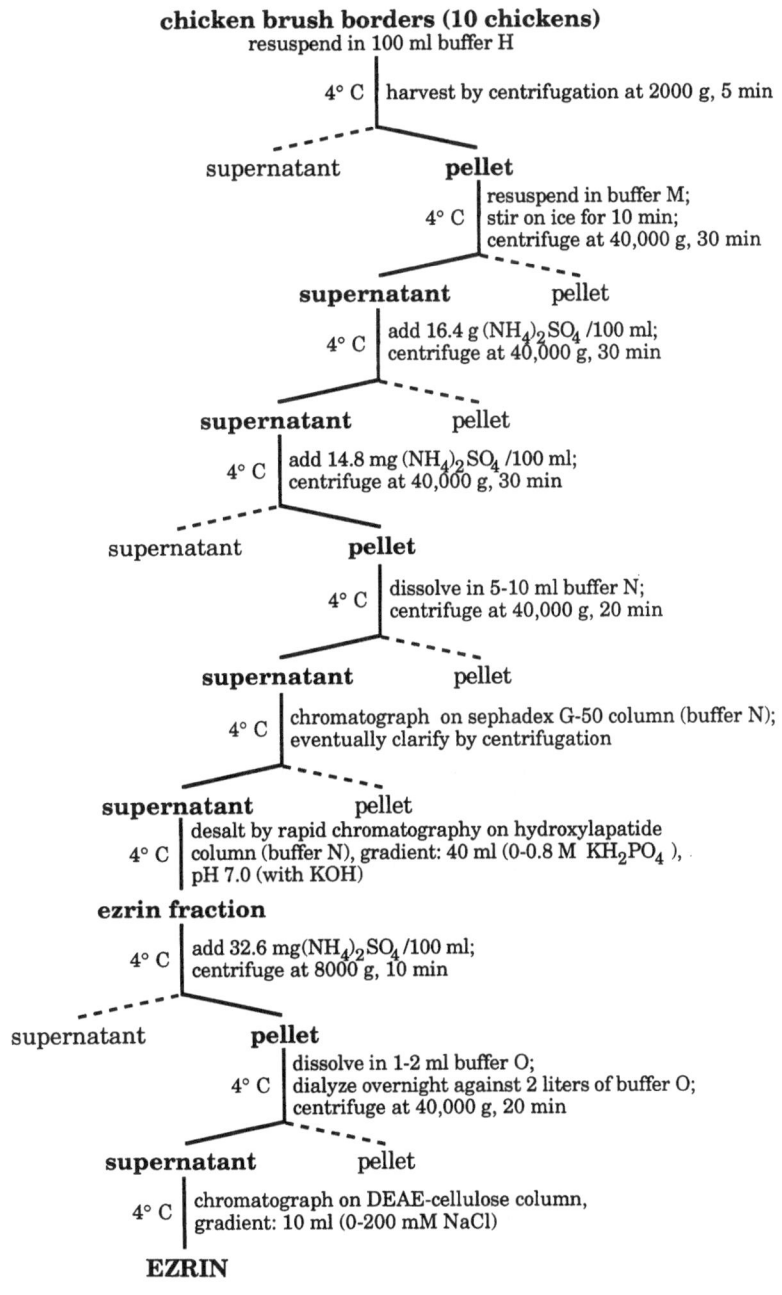

4° C | harvest by centrifugation at 2000 g, 5 min

supernatant   **pellet**

4° C | resuspend in buffer M;
stir on ice for 10 min;
centrifuge at 40,000 g, 30 min

supernatant   pellet

4° C | add 16.4 g $(NH_4)_2SO_4$ /100 ml;
centrifuge at 40,000 g, 30 min

supernatant   pellet

4° C | add 14.8 mg $(NH_4)_2SO_4$ /100 ml;
centrifuge at 40,000 g, 30 min

supernatant   **pellet**

4° C | dissolve in 5-10 ml buffer N;
centrifuge at 40,000 g, 20 min

supernatant   pellet

4° C | chromatograph on sephadex G-50 column (buffer N);
eventually clarify by centrifugation

supernatant   pellet

4° C | desalt by rapid chromatography on hydroxylapatide
column (buffer N), gradient: 40 ml (0-0.8 M KH_2PO_4),
pH 7.0 (with KOH)

ezrin fraction

4° C | add 32.6 mg$(NH_4)_2SO_4$ /100 ml;
centrifuge at 8000 g, 10 min

supernatant   **pellet**

4° C | dissolve in 1-2 ml buffer O;
dialyze overnight against 2 liters of buffer O;
centrifuge at 40,000 g, 20 min

supernatant   pellet

4° C | chromatograph on DEAE-cellulose column,
gradient: 10 ml (0-200 mM NaCl)

EZRIN

FASCIN

A 58 kDa F-actin bundling protein presumably responsible for arranging actin in filopodia.

Source:	sea urchin eggs
Equipment:	• hand centrifuge
	• Dounce homogenizer
	• centrifuge
	• water bath
	• A5m argarose column (1 x 25 cm)
	• DEAE column (1 x 3 cm)
Chemicals:	KCl, NaCl, EDTA, glycerol, PIPES, EGTA, ATP, argarose, KI, DEAE-sephacel

Have ready:

isotonic KCl: 0.5 M

filtered seawater: 19:1:19, 0.5 M NaCl, 0.5 M KCl, 2 mM EDTA (Millipore 0.45 μm)

extraction buffer: 0.9 M glycerol, 0.1 M PIPES, 5 mM EGTA, pH 6.9

dialysis buffer: 10 mM PIPES, 0.1 mM EGTA, 0.1 mM ATP, pH 6.9

gelwashing buffer: 0.1 M KCl, 10 mM PIPES, 0.1 mM EGTA, pH 6.8

KI-buffer: 1.5 M KJ, 10 mM ATP, 1 mM EGTA, 10 mM PIPES, pH 6.8

argarose column buffer: 0.6 M KCl, 10 mM PIPES, pH 6.8

DEAE column buffer: 10 mM PIPES, 0.1 mM EGTA, pH 6.8

Reference: Bryan, J. (1986) Methods Enzymol. 134, 13-23

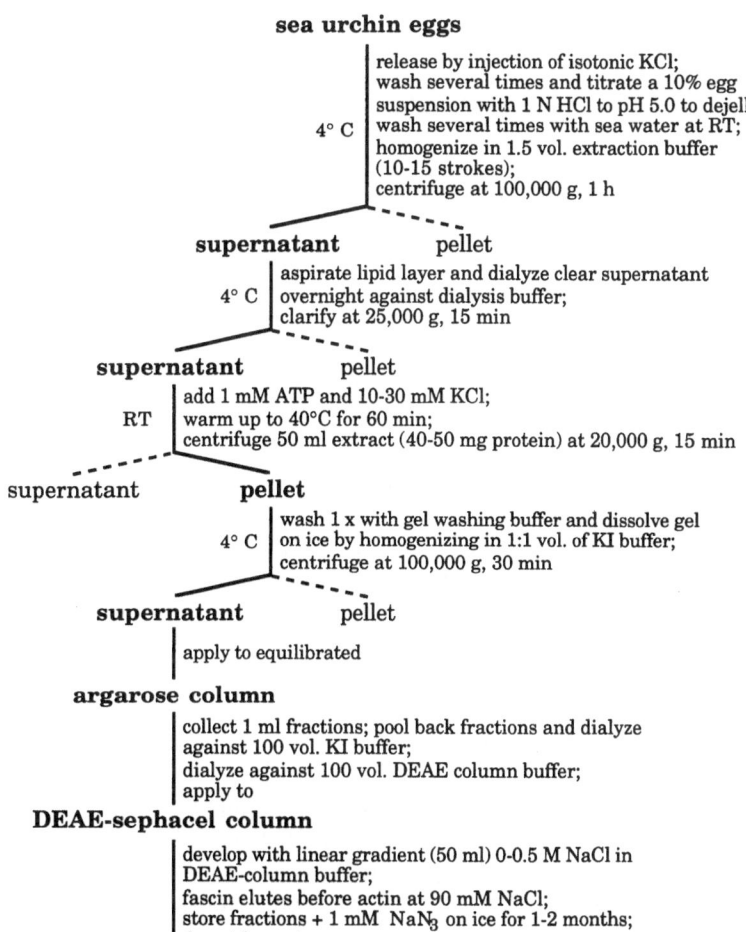

sea urchin eggs

4° C | release by injection of isotonic KCl;
wash several times and titrate a 10% egg
suspension with 1 N HCl to pH 5.0 to dejell;
wash several times with sea water at RT;
homogenize in 1.5 vol. extraction buffer
(10-15 strokes);
centrifuge at 100,000 g, 1 h

supernatant   pellet

4° C | aspirate lipid layer and dialyze clear supernatant
overnight against dialysis buffer;
clarify at 25,000 g, 15 min

supernatant   pellet

RT | add 1 mM ATP and 10-30 mM KCl;
warm up to 40°C for 60 min;
centrifuge 50 ml extract (40-50 mg protein) at 20,000 g, 15 min

supernatant   **pellet**

4° C | wash 1 x with gel washing buffer and dissolve gel
on ice by homogenizing in 1:1 vol. of KI buffer;
centrifuge at 100,000 g, 30 min

supernatant   pellet

apply to equilibrated

argarose column

collect 1 ml fractions; pool back fractions and dialyze
against 100 vol. KI buffer;
dialyze against 100 vol. DEAE column buffer;
apply to

DEAE-sephacel column

develop with linear gradient (50 ml) 0-0.5 M NaCl in
DEAE-column buffer;
fascin elutes before actin at 90 mM NaCl;
store fractions + 1 mM NaN$_3$ on ice for 1-2 months;
do not freeze !

FASCIN

FILAMIN

Filamin (ABP-280), a 280 kDa homodimeric phosphoprotein that crosslinks actin filaments. Molecular contour length: 160 nm, 2647 amino acids, ubiquitous.

Source: chicken gizzard smooth muscle

Equipment:
- mincer
- Waring blender
- glass-teflon-homogenizer with pestle
- sepharose 6B column (2.5 x 90 cm)
- DEAE-cellulose column (2.5 x 20 cm)
- Amicon cell
- electrophoretic gels
- centrifuge

Chemicals: KCl, Na_2ATP, EDTA, DTT, $(NH_4)_2SO_4$ (up)
Trizma Base, NaN_3
protease inhibitors: PMSF
column materials: sepharose 6B, DEAE-cellulose

Have ready: **buffer A (low salt extraction):** 5 mM EDTA, 5 mM DTT, 0.5 mM PMSF, pH 7.0 (600 ml)

buffer B (high salt extraction): 50 mM Tris-HCl, 0.5 mM PMSF, pH 7.5 (500 ml)

buffer C (dialysis buffer): 50 mM Tris-HCl, 0.1 M KCl, 1 mM EDTA, 0.5 mM DTT, pH 7.5 (40 l)

sepharose-column buffer: 50 mM Tris-HCl, 0.5 mM EDTA, 0.5 mM DTT, 0.05 % NaN_3, pH 7.5 (5 l)

DEAE-cellulose column buffer: 20 mM Tris-HCl, 0.5 mM EDTA, 0.5 mM DTT, pH 7.5 (10 liters)

Reference: Hartwig, J. and Stossel, T. (1981)
J.Mol.Biol. 145, 563-581.

chicken gizzard smooth muscle (~500 g)

4°C │ mince with scissors; extract with buffer A;
homogenize with Waring blender, 1 min;
centrifuge at 27,000 g, 30 min

supernatant
low salt extract   **pellet**

4°C │ extract with buffer B;
homogenize with Waring blender, 30 sec;
centrifuge at 27,000 g, 30 min

supernatant   pellet
high salt extract

4°C │ dialyze against buffer C, 8 h;
centrifuge at 27,000 g, 20 min

supernatant   pellet

4°C │ add 20.5 g $(NH_4)_2SO_4$ / 100 ml;
centrifuge at 24,000 g, 20 min

supernatant   **pellet**

4°C │ disperse in 25 ml+200 ml buffer C;
dialyze against buffer C, 10 h;
centrifuge at 30,000 g, 2 h

supernatant   pellet

4°C │ add 20.5 g $(NH_4)_2SO_4$ / 100 ml;
centrifuge at 24,000 g, 20 min

supernatant   **pellet**

4°C │ dissolve in 20 ml sepharose buffer;
centrifuge at 100,000 g, 30 min

supernatant   pellet

4°C │ chromatograph on sepharose 6 B

filamin fractions

4°C │ chromatograph on DEAE-cellulose,
gradient: 400 ml (0-0.5 M KCl);
pool filamin fractions

FILAMIN
(~50 mg / 100 g gizzards)

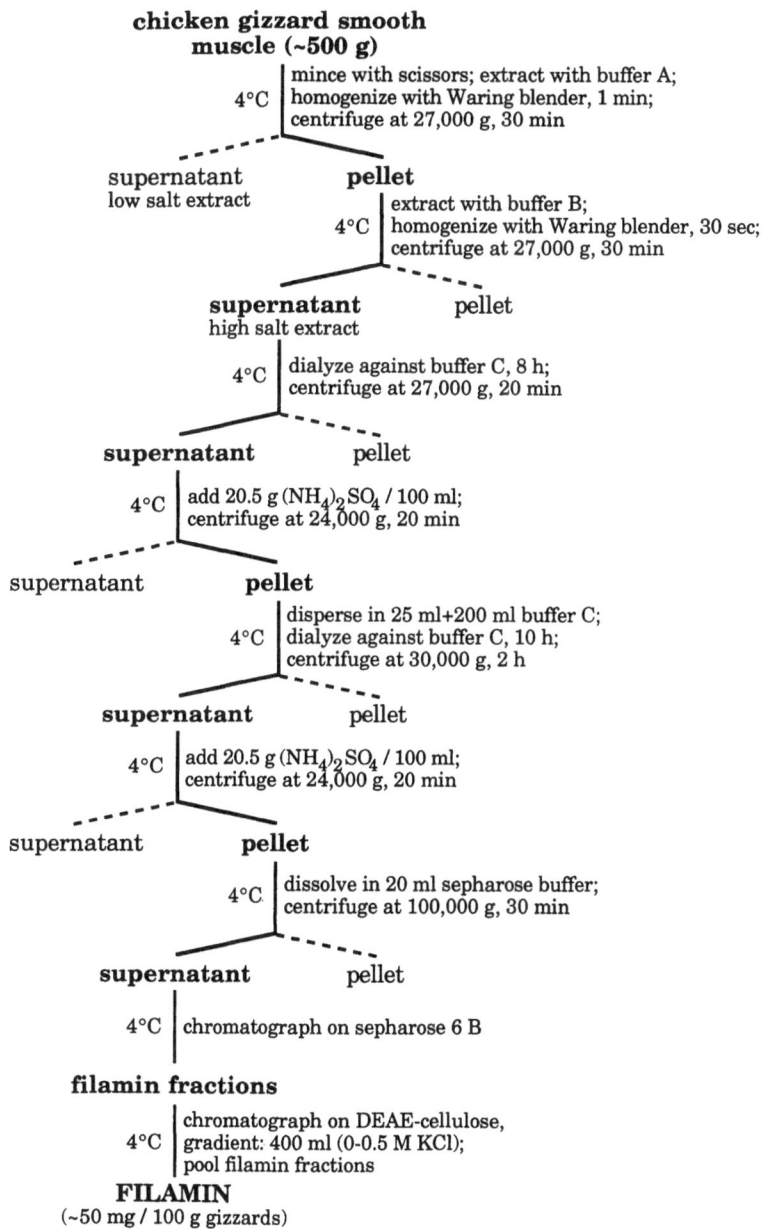

FIMBRIN

A monomeric F-actin bundling protein of 68 kDa found in microvilli and membrane ruffles of non muscle cells

Source:	brush borders from chicken intestines
Equipment:	• DNase I immobilized on sepharose 4B column
	• sephacryl S-200 column (200 ml)
	• DEAE-cellulose column (5 ml)
	• DNase I immobilized on Affi-Gel-Blue column
	• electrophoretic gels
	• centrifuge
Chemicals:	$CaCl_2$, NaN_3, Trizma-Base, KCl, DTT, EGTA, $MgCl_2$, imidazole, $(NH_4)_2SO_4$ (up)
	protease inhibitors: PMSF, benzamidine
	column material: sephacryl S-200, DEAE-cellulose
Have ready: HCl,	**buffer A:** 5 mM $CaCl_2$, 1 mM NaN_3, 10 mM Tris-pH 7.5
	buffer H: 75 mM KCl, 1 mM EGTA, 0.1 mM $MgCl_2$, 0.25 mM PMSF, 0.5 mM benzamidine, 10 mM imidazole-HCl, pH 7.3
mM	**buffer I:** 0.2 M KCl, 1 mM $CaCl_2$, 1 mM DTT, 10 imidazole-HCl, pH 7.3
	buffer L: 50 mM NaCl, 1 mM DTT, 10 mM imidazole-HCl, pH 6,8
Reference:	Bretscher, A. (1986) Methods Enzymol. 134, 32-34
	Gen-Bank: X5262

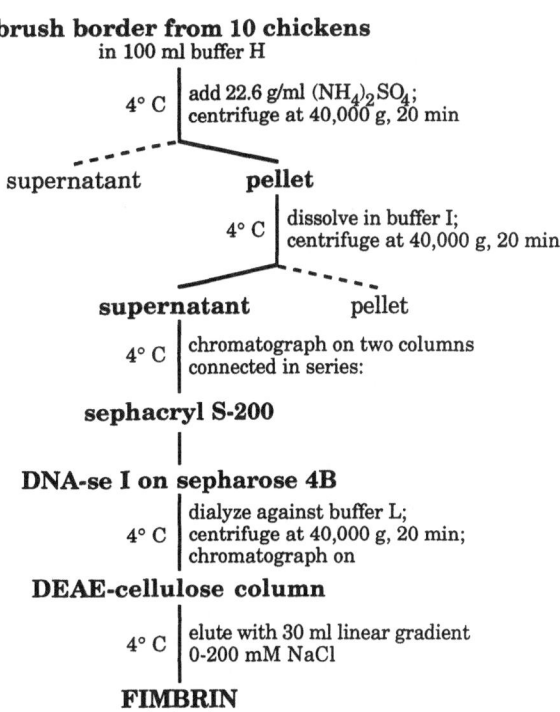

brush border from 10 chickens
in 100 ml buffer H

4° C | add 22.6 g/ml $(NH_4)_2SO_4$;
centrifuge at 40,000 g, 20 min

supernatant   **pellet**

4° C | dissolve in buffer I;
centrifuge at 40,000 g, 20 min

supernatant   pellet

4° C | chromatograph on two columns
connected in series:

sephacryl S-200

DNA-se I on sepharose 4B

4° C | dialyze against buffer L;
centrifuge at 40,000 g, 20 min;
chromatograph on

DEAE-cellulose column

4° C | elute with 30 ml linear gradient
0-200 mM NaCl

FIMBRIN
(~5 mg)

FODRIN

Fodrin (non-erythroid α-spectrin) is a 284 kDa protein consisting of 2,472 amino acids. Mostly α-helical. A lipid and actin binding linker protein of plasma membranes.

Source: bovine, porcine, and guinea pig brain

Equipment:
- Hamilton-beach blender
- centrifuge
- Brinkmann polytron homogenizer
- sepharose Cl-4B column (2.5 x 90 cm)
- electrophoretic gels
- electrophoretic gradient gel
- microtip probe sonicator
- Amicon cell PM 10
- sucrose gradient (5-20%)
- Beckman SW41 Ti rotor

Chemicals: imidazole, EDTA, NaN_3, EGTA, Trizma-Base, KCl, $MgCl_2$, $CaCl_2$, DTT, KI, Na_2HPO_4, $(NH_4)_2SO_4$
protease inhibitors: benzamidine, PMSF
column materials: sepharose CL-4B

Have ready: **homogenization buffer:** 10 mM imidazole-HCl, 5 mM EDTA, 3 mM NaN_3, 1 mM EGTA, pH 7.3

extraction buffer: 10 mM Tris-HCl, 50 mM KCl, 3 mM NaN_3, 1 mM $MgCl_2$, 0.1 mM $CaCl_2$, 0.1 mM DTT, pH 8.2

TKE/KI: 10 mM Tris-HCl, 700 mM KI, 0.5 mM EDTA, 0.1 mM DTT

TKE/KCl: 10 mM Tris-HCl, 700 mM KCl, 0.5 mM EDTA, 0.1 mM DTT

precipitation buffer: 10 mM Tris-HCl, 3 mM NaN_3, 0.1 mM $CaCl_2$, 0.1 mM DTT, pH 8.2

storage buffer: 1 mM Na_2HPO_4, 3 mM NaN_3, 0.1 mM DTT, pH 8.0

Reference: Cheney, R., Levine, J. and Willard, M. (1986) Methods Enzymol. 134, 42-54
Gen-Bank: J05243 (human brain, fibroblast α-spectrin)

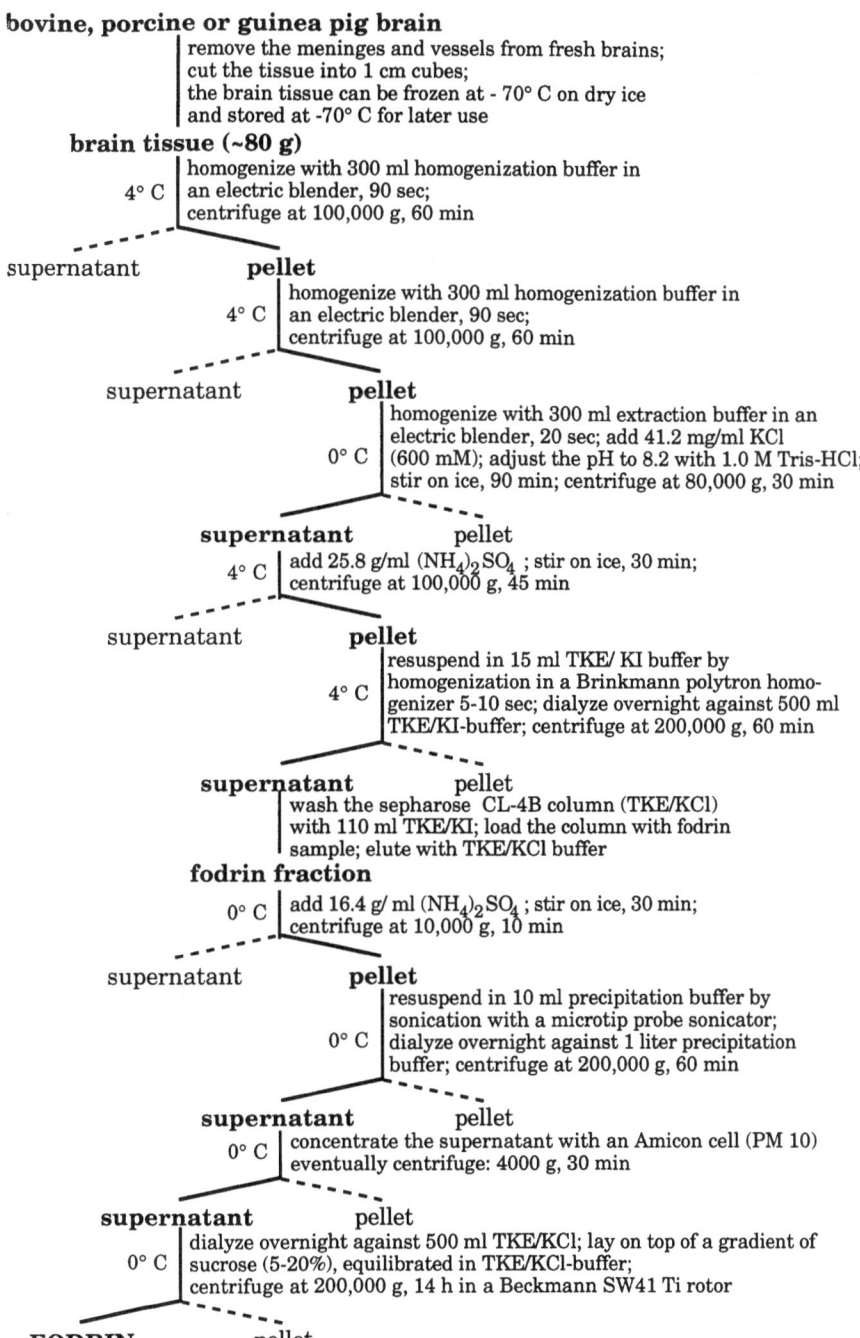

bovine, porcine or guinea pig brain
  remove the meninges and vessels from fresh brains;
  cut the tissue into 1 cm cubes;
  the brain tissue can be frozen at - 70° C on dry ice
  and stored at -70° C for later use

brain tissue (~80 g)
  4° C | homogenize with 300 ml homogenization buffer in
        an electric blender, 90 sec;
        centrifuge at 100,000 g, 60 min

supernatant   **pellet**
  4° C | homogenize with 300 ml homogenization buffer in
        an electric blender, 90 sec;
        centrifuge at 100,000 g, 60 min

supernatant   **pellet**
  0° C | homogenize with 300 ml extraction buffer in an
        electric blender, 20 sec; add 41.2 mg/ml KCl
        (600 mM); adjust the pH to 8.2 with 1.0 M Tris-HCl;
        stir on ice, 90 min; centrifuge at 80,000 g, 30 min

supernatant   pellet
  4° C | add 25.8 g/ml $(NH_4)_2SO_4$; stir on ice, 30 min;
        centrifuge at 100,000 g, 45 min

supernatant   **pellet**
  4° C | resuspend in 15 ml TKE/ KI buffer by
        homogenization in a Brinkmann polytron homo-
        genizer 5-10 sec; dialyze overnight against 500 ml
        TKE/KI-buffer; centrifuge at 200,000 g, 60 min

supernatant   pellet
        wash the sepharose CL-4B column (TKE/KCl)
        with 110 ml TKE/KI; load the column with fodrin
        sample; elute with TKE/KCl buffer

fodrin fraction
  0° C | add 16.4 g/ ml $(NH_4)_2SO_4$; stir on ice, 30 min;
        centrifuge at 10,000 g, 10 min

supernatant   **pellet**
  0° C | resuspend in 10 ml precipitation buffer by
        sonication with a microtip probe sonicator;
        dialyze overnight against 1 liter precipitation
        buffer; centrifuge at 200,000 g, 60 min

supernatant   pellet
  0° C | concentrate the supernatant with an Amicon cell (PM 10)
        eventually centrifuge: 4000 g, 30 min

supernatant   pellet
  0° C | dialyze overnight against 500 ml TKE/KCl; lay on top of a gradient of
        sucrose (5-20%), equilibrated in TKE/KCl-buffer;
        centrifuge at 200,000 g, 14 h in a Beckmann SW41 Ti rotor

FODRIN   pellet

83

gCAP 39

A Ca⁺⁺ activated and PI-inhibited actin filament capping protein of ~40 kDa abundant in macrophages.

Source:	alveolar macrophages
Equipment:	• Dounce homogenizer
	• water bath
	• DEAE-sepharose CL-6B column (2.5 x 30 cm)
	• vacuum dialysis, Schleicher&Schnell UH 100/10
	• Protein PAK 125 and PAK 300 silica high pressure gel filtration columns (Waters Ass.)
Chemicals:	sucrose, EDTA, DTT, ATP, Tris-maleate, imidazole, EGTA
Have ready:	**homogenizing buffer:** 0.34 M sucrose, 10 mM DTT, 0.5 mM ATP, 1 mM EDTA, 20 mM Tris-maleate, pH 7.0
	buffer B: 10 mM imidazole, 1 mM DTT, 0.5 mM ATP, 1 mM EGTA, pH 7.8
Reference:	Southwick, F.S. and DiNubile, N.J. (1986) J. Biol. Chem. 261, 14191-14195
	Gen-Bank: M38463 (mouse)

macrophages (10-15 ml packed cells)

4° C | grind with Dounce homogenizer in
homogenizing buffer (2 vol.);
centrifuge at 100,000 g, 1 h

supernatant   pellet

4° C | make 0.1 M with 3 M KCl and
incubate at 25° C for 1.5 h;
centrifuge at 12,000 g, 10 min

supernatant   pellet

dialyze overnight against 400 vol. buffer B;
apply to

DEAE-sepharose column

equilibrate and wash with buffer B; elute by a linear
gradient 0.0-0.45 M KCl (1 l total) in buffer B;
concentrate approx. 50-fold by vacuum dialysis (UH
100/10) in buffer B + 0.1 M KCl and apply
sequentially to two of each columns in series

PAK 125 (2-80,000 MW)

PAK 300 (10-400,000 MW)

elute with buffer B + 0.1 M KCl, pH 7.3 at 0.5 ml/min

gCAP 39

GELSOLIN

A widespread Ca^{++}- and phosphoinositide-regulated actin filament severing protein of 80 kDa molecular weight.

Source: bovine plasma, or porcine plasma

Equipment:
- DEAE-cellulose column (5 x 50 cm)
- DEAE-cellulose column (2.5 x 20 cm)
- electrophoretic gels
- falling ball viscometry

Chemicals: Trizma Base, CaCl$_2$, EGTA, NaCl
column materials: DEAE-cellulose

Have ready: **DEAE-column buffer I:** 25 mM Tris-HCl, 0.5 mM CaCl$_2$, pH 7.5 (6 liters)

DEAE-column buffer II: 25 mM Tris-HCl 0.1 mM EGTA, pH 7.5 (3 liters)

Reference: Chaponnier, C., Janmey, P.A., and Yin, H. (1986) J. Cell Biol. 103, 1473-1481

bovine plasma (500 ml)

RT | dialyze against DEAE column-buffer, 12 h;
   | centrifuge at 12,000 g, 30 min

supernatant   pellet

RT | chromatograph at RT on

DEAE column

4° C | (flow rate 60 ml/h);
     | pool the gelsolin fractions;
     | cool to 4° C;
     | divide the gelsolin fractions in two;
     | adjust pH to 7.5 and add 5 mM EGTA final conc.;
     | chromatograph on

DEAE column

| gradient: 500 ml (0-0.5 M NaCl), (flow rate 30 ml/h)

GELSOLIN

HISACTOPHILIN

A 17 kDa, pH-sensitive actin binding protein enriched in subplasma
membrane areas in Dictyostelium amoeba, carrying 31 histidines out
of 118 amino acids.

Source:	Dictyostelium discoideum
Equipment:	• Parr bomb
	• centrifuge
	• DEAE (DE-52) column (5 x 40 cm)
	• sepharose 6B-CL-column (5 x 95 cm)
	• DEAE (DE-52) column (1.5 x 1 cm)
Chemicals:	Tris-HCl, DTT, ATP, EGTA, sucrose, NaCl, NaN$_3$, (NH$_4$)$_2$SO$_4$ solid
	protease inhibitors: benzamidine, o-phenanthroline, PMSF, N-carbobenzoxy-phenylalanine, N$^+$-p-tosyl-L-arginine methylester, N-tosyl-L-phenylalanine chloromethylketone, pepstatin, leupeptine, trypsin/aprotinin
Have ready:	**homogenization buffer:** 30 mM Tris-HCl, 2 mM DTT, 4 mM EGTA, 0.2 mM ATP, 30% w/v sucrose, 5 mM benzamidine, 5 mM o-phenanthroline, 0.5 mM PMSF, 2 mM N-carbobenzoxy-phenylalanine, 0.1 mg/ml N$^+$-p-tosyl-L-arginine methylester, 0.1 mg/ml N-tosyl-L-phenylalanine chloromethyl-ketone, 2 µg/ml pepstatin, 5 µg/ml leupeptine, 0.15 IU trypsin /ml aprotinin, pH 7.6
	DEAE column buffer: 10 mM Tris-HCl, 1 mM EGTA, 1 mM DTT, 0.02% NaN$_3$, 1 mM benzamidine, pH 7.6
Reference:	Scheel et al. (1989) J. Biol. Chem. 264, 2832-2839 Gen-Bank: EMBL J04472

Dictyostelium cells (~600 g)

4° C │ homogenize in Parr bomb;
     │ centrifuge at 100,000 g, 2 h

supernatant   pellet

│ load onto

DE-52 column

│ elute with linear gradient 0-350 mM NaCl (2 x 750 ml);
│ collect fractions between 40-100 mM NaCl;
│ concentrate with $(NH_4)_2SO_4$ (55%);
│ redissolve in 18 ml portions in DEAE-buffer + 0.2 M NaCl and
│ apply each fraction on

sepharose 6B-CL-column

│ hisactophilin elutes after the indended volume from
│ gel-filtration columns;
│ concentrate on small DEAE column if necessary and
│ step-elute with DEAE-buffer incl. 0.2 M NaCl

HISACTOPHILIN

IAP (Inhibitor of Actin Polymerization)

A 25 kDa low molecular weight heat shock protein from muscle. 193 amino acids, probably a non-nucleating barbed-end F-actin capping protein. Different from tensin and insertin.

Source: turkey gizzard smooth muscle

Equipment:
- DEAE-cellulose (DE-52) column (2.6 x 23 cm)
- hydroxylapatide column (1.3 x 11 cm) Bio-gel HTP

Chemicals: Tris-HCl, NaCl, 2-mercaptoethanol, EDTA
protease inhibitors: PMSF

Have ready: **buffer A:** 20 mM Tris-HCl, 10 mM NaCl, 15 mM 2-mercaptoethanol, 0.1 mM EDTA, 0.1 mM PMSF, pH 7.6

phosphate buffer, 50 mM, pH 7.0

Reference:
Miron et al. (1988) Europ. J. Biochem. 187, 543-553
Miron et al. (1991) J. Cell. Biol. 114, 255-261
Gen-Bank: X59541

turkey gizzard (~150 g)

> follow vinculin preparation (p.   ) up
> to crude vinculin fractions; dialyze against buffer A and
> apply crude IAP-extract (~175 ml, 1 g protein) to

DEAE-cellulose column

> preequilibrate with buffer A;
> wash with 7.5 bed volume buffer A;
> elute with linear NaCl gradient (0-0.25 M in buffer A, 2 x 400 ml);
> collect 4.9 ml fractions; pool active fractions; dialyze against
> buffer A; adjust to 15 mM phosphate and load onto

hydroxylapatide column

> preequilibrate with 50 mM phosphate buffer;
> load 24 mg total protein; wash;
> elute with linear phosphate gradient (50-250 mM, pH 7.0,
> 2 x 100 ml); collect 2.8 ml fractions

IAP

INSERTIN

A vinculin associated 30 kDa protein which binds to the barbed ends of actin filaments but unlike capping proteins does not completely prevent polymerization.

Source:	chicken gizzard smooth muscle
Equipment:	• Waring blender • centrifuge • DEAE cellulose column (1.2 x 20 cm) • DEAE sepharose CL-6B column (1.2 x 20 cm) • sephacryl S-300 column (2.5 x 80 cm) • hydroxylapatide column (1.5 x 12 cm)
Chemicals:	Tris-HCl, EGTA, acetic acid, $MgCl_2$, solid $(NH_4)_2SO_4$, EDTA, Tris-acetate, 2-mercaptoethanol, NaCl, KH_2PO_4, NaOH, polyethylene glycol (PEG) **protease inhibitors:** PMSF
Have ready:	**buffer A:** 2 mM Tris-HCl, 1 mM EGTA, 0.5 mM PMSF, pH 9.0
	buffer B: 20 mM NaCl, 0.1 M EDTA, 15 mM 2-mercaptoethanol, 20 mM Tris-acetate, pH 7.6
acetate,	**buffer C:** 5 mM NaCl, 0.25 mM EDTA, 0.1 mM PMSF, 5 mM 2-mercaptoethanol, 20 mM Tris- pH 7.6
	buffer D: 100 mM NaCl, 10 mM KH_2PO_4-NaOH, pH 7.5
Reference:	Ruhnau, K., Gaertner, A. and Wegner, A. (1989) J. Mol. Biol. 210, 141-148

92

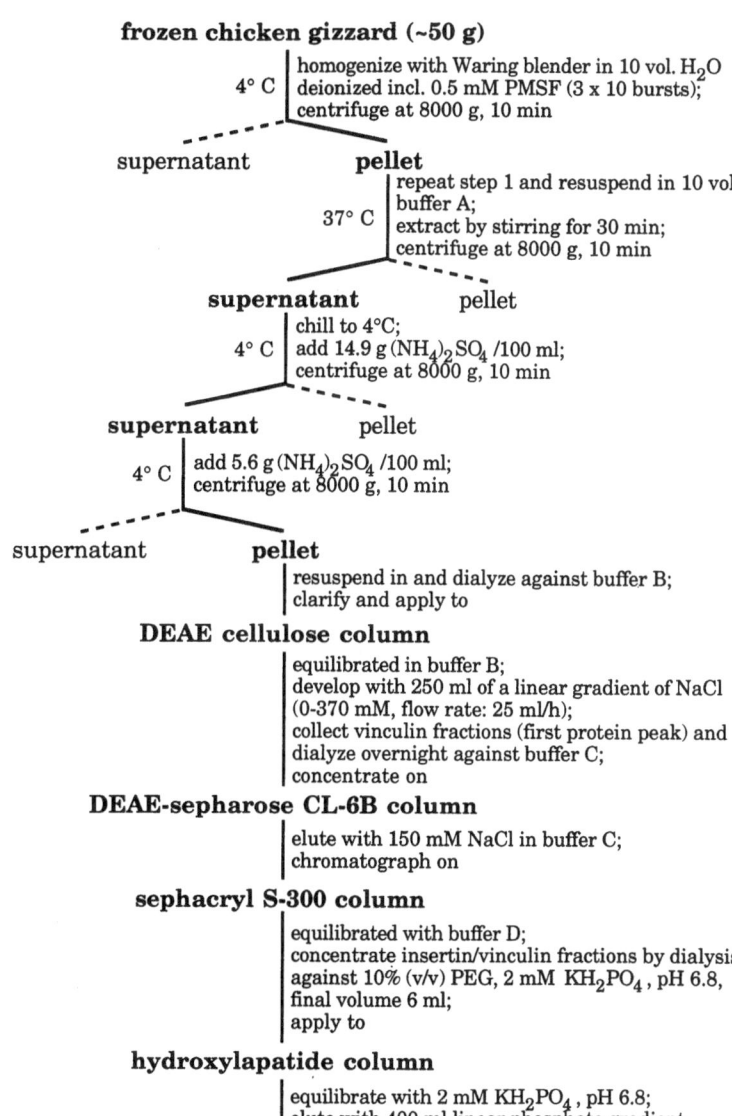

frozen chicken gizzard (~50 g)

> 4° C | homogenize with Waring blender in 10 vol. H_2O deionized incl. 0.5 mM PMSF (3 x 10 bursts); centrifuge at 8000 g, 10 min

supernatant   **pellet**

> 37° C | repeat step 1 and resuspend in 10 vol. buffer A; extract by stirring for 30 min; centrifuge at 8000 g, 10 min

supernatant   pellet

> 4° C | chill to 4°C; add 14.9 g $(NH_4)_2SO_4$ /100 ml; centrifuge at 8000 g, 10 min

supernatant   pellet

> 4° C | add 5.6 g $(NH_4)_2SO_4$ /100 ml; centrifuge at 8000 g, 10 min

supernatant   **pellet**

> resuspend in and dialyze against buffer B; clarify and apply to

DEAE cellulose column

> equilibrated in buffer B; develop with 250 ml of a linear gradient of NaCl (0-370 mM, flow rate: 25 ml/h); collect vinculin fractions (first protein peak) and dialyze overnight against buffer C; concentrate on

DEAE-sepharose CL-6B column

> elute with 150 mM NaCl in buffer C; chromatograph on

sephacryl S-300 column

> equilibrated with buffer D; concentrate insertin/vinculin fractions by dialysis against 10% (v/v) PEG, 2 mM KH_2PO_4, pH 6.8, final volume 6 ml; apply to

hydroxylapatide column

> equilibrate with 2 mM KH_2PO_4, pH 6.8; elute with 400 ml linear phosphate gradient (2 mM-200 mM phosphate); collect 3 ml fractions

INSERTIN

M-LINE PROTEIN
(MM-Creatine Kinase)

MM-CK- the M-type isoprotein of creatine kinase, in conjunction with myomesin the only so far identified true M-line protein, function unknown.

Source:	chicken breast muscle
Equipment:	• mixer • Millipore filter • centrifuge • nylon cloth
Chemicals:	Tris-HCl, KCl, EGTA, EDTA, DTT, CH_3COOH, $(NH_4)_2SO_4$, DEAE-cellulose (DE-52), HCl, NaOH
Have ready:	**washing buffer:** 0.1 M KCl, 1 mM EGTA, 5 mM EDTA, 1 mM DTT, pH 7.0 (with NaOH) **extraction buffer:** 5 mM Tris-HCl, 1 mM DTT, pH 7.7
Reference:	Eppenberger, H.M. and Strehler, E. (1982) Methods Enzymol. 85, 139-143

94

breast muscle (~400 g)

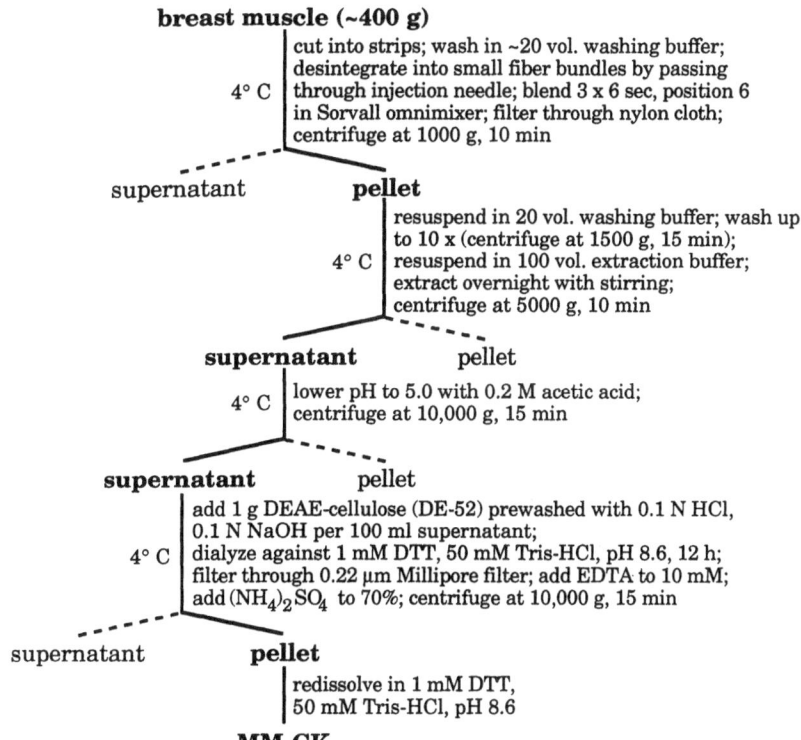

4° C | cut into strips; wash in ~20 vol. washing buffer; desintegrate into small fiber bundles by passing through injection needle; blend 3 x 6 sec, position 6 in Sorvall omnimixer; filter through nylon cloth; centrifuge at 1000 g, 10 min

supernatant   **pellet**

4° C | resuspend in 20 vol. washing buffer; wash up to 10 x (centrifuge at 1500 g, 15 min); resuspend in 100 vol. extraction buffer; extract overnight with stirring; centrifuge at 5000 g, 10 min

supernatant   pellet

4° C | lower pH to 5.0 with 0.2 M acetic acid; centrifuge at 10,000 g, 15 min

supernatant   pellet

4° C | add 1 g DEAE-cellulose (DE-52) prewashed with 0.1 N HCl, 0.1 N NaOH per 100 ml supernatant; dialyze against 1 mM DTT, 50 mM Tris-HCl, pH 8.6, 12 h; filter through 0.22 μm Millipore filter; add EDTA to 10 mM; add $(NH_4)_2SO_4$ to 70%; centrifuge at 10,000 g, 15 min

supernatant   **pellet**

redissolve in 1 mM DTT, 50 mM Tris-HCl, pH 8.6

MM-CK

MARCKS

A 27-32 kDa myristoylated actin binding phosphoprotein with an
apparent MW of 87 kDa on SDS gels. Substrate for protein kinase C
and colocalizing with vinculin and talin at adhesion sites.

Source:	rat brain
Equipment:	• Waring blender
	• centrifuge
	• water bath with stirrer
	• mono-Q-column (Pharmacia), (0.5 x 5 cm)
	• reversed phase column (Pharmacia) C-8, Pro RCP, 15 μm (1 x 10 cm)
	• reversed phase column (Pharmacia) C-8, Pro RCP, 5 μm (0.5 x 10 cm)
Chemicals:	Tris-HCl, EGTA, DTT, $(NH_4)_2SO_4$ solid, NaCl, trifluoroacetic acid, acetonitrile
	protease inhibitors: PMSF, leupeptin
Have ready:	**buffer A:** 20 mM Tris-HCl, 5 mM EGTA, 5 mM DTT, 12 μM leupeptin, 0.1 mM PMSF, pH 7.4
Reference:	Patel, J. and Kligman, D. (1987) J. Biol. Chem 262, 16686-16691
	Gen-Bank: M23738 (bovine), M60474 (mouse)

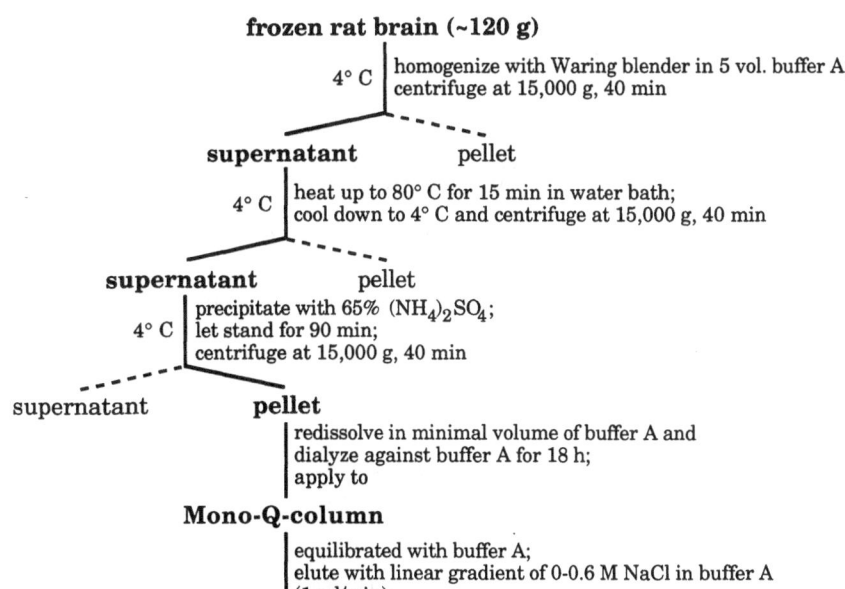

frozen rat brain (~120 g)

4° C | homogenize with Waring blender in 5 vol. buffer A;
centrifuge at 15,000 g, 40 min

supernatant   pellet

4° C | heat up to 80° C for 15 min in water bath;
cool down to 4° C and centrifuge at 15,000 g, 40 min

supernatant   pellet

4° C | precipitate with 65% $(NH_4)_2SO_4$;
let stand for 90 min;
centrifuge at 15,000 g, 40 min

supernatant   **pellet**

redissolve in minimal volume of buffer A and
dialyze against buffer A for 18 h;
apply to

Mono-Q-column

equilibrated with buffer A;
elute with linear gradient of 0-0.6 M NaCl in buffer A
(1 ml/min);
collect 0.5 ml fractions;
adjust to 0.1% trifluoracetic acid and apply to

reversed phase column (ø 15 μm)

equilibrate column with 0.1% trifluoracetic acid in water;
elute with linear gradient (0-60%) acetonitrile in 0.1%
trifluoracetic acid; collect 0.5 ml fractions;
dialyze pooled 87 kDa protein against 0.1% trifluoracetic
acid for 18 h; apply to

reversed phase column (ø 5 μm)

elute with 5 ml linear gradient 0-30% acetonitrile in
0.1% trifluoracetic acid followed by isocratic elution
(5 ml) with 30% acetonitrile in 0.1% trifluoracetic
acid (1 ml/min)

MARCKS

METAVINCULIN

| **A 152 kDa protein largely homologous with vinculin.** |

Source: trimmed glycerinated chicken gizzards, stored at -20° C

Equipment:
- DEAE-cellulose column (5.3 x 18.5 cm)
- hydroxylapatide column (2.1 x 7.9 cm)
- HPLC with mono Q anion exchange resin
- Acrodisc 0.2 mm filters
- electrophoretic gels
- centrifuge

Chemicals: $(NH_4)_2SO_4$ (up), DMSO, Trizma-Base, EDTA, EGTA, DTT, NaN_3, NaH_2PO_4, NaCl, KCl
protease inhibitors: trasylol, leupeptin, antipain, benzamidine, chymostatin, pepstatin
column materials: DEAE-cellulose, hydroxylapatide

Have ready: **protease inhibitor cocktail I (PIC I):** 1 mg/ml leupeptin, 2 mg/ml antipain, 10 U/ml trasylol, 10 mg/ml benzamidine, all components are dissolved in trasylol

protease inhibitor cocktail II (PIC II): 1 mg/ml chymostatin, 1 mg/ml pepstatin, both are dissolved together in DMSO

extraction buffer: 0.6 M Tris-HCl, 1 mM EDTA, 1 mM EGTA, 10 U/ml trasylol, 1 ml PIC I, 1 ml PIC II, pH 7.4 (1 liter)

Tris-HCl buffer: 50 mM Tris-HCl, pH 7.4 (20 liters)

DEAE-column buffer: 50 mM Tris-HCl, 1 mM EDTA, 1 mM DTT, 10 U/ml trasylol, 0.02% NaN_3 (w/v), pH 7.4 at 24° C (24 liters)

hydroxylapatide-column buffer: 10 mM NaH_2PO_4, 0.15 M NaCl, 0.02% NaN_3 (w/v), pH 7.3 (10 liters)

HPLC-column buffer: this buffer is identical to DEAE-column buffer (see above)

Reference: Siliciano, J.D. and Craig (1982) Nature 300, 533-535
O'Halloran, T., Molony, L. and Burridge, K. (1986) Methods Enzymol. 134, 69-77

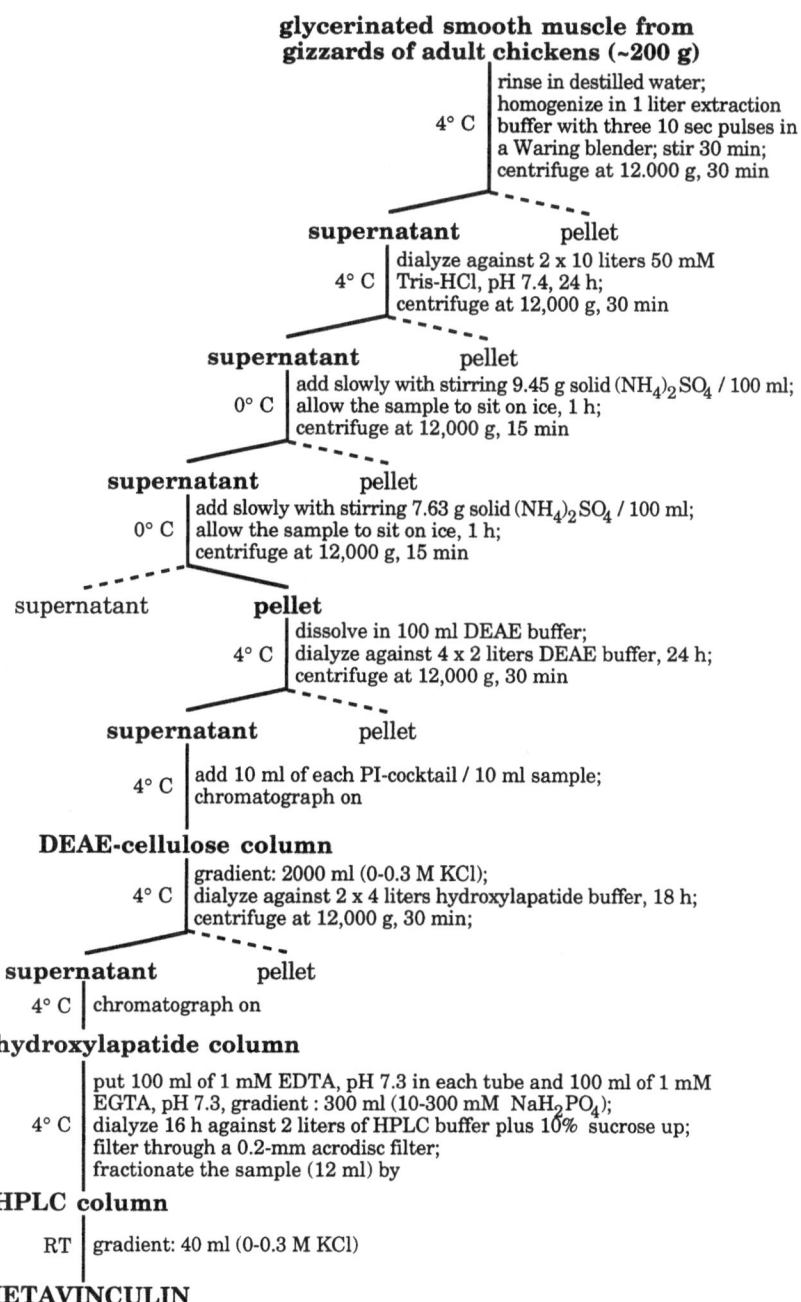

glycerinated smooth muscle from gizzards of adult chickens (~200 g)

4° C | rinse in destilled water; homogenize in 1 liter extraction buffer with three 10 sec pulses in a Waring blender; stir 30 min; centrifuge at 12.000 g, 30 min

supernatant   pellet

4° C | dialyze against 2 x 10 liters 50 mM Tris-HCl, pH 7.4, 24 h; centrifuge at 12,000 g, 30 min

supernatant   pellet

0° C | add slowly with stirring 9.45 g solid $(NH_4)_2SO_4$ / 100 ml; allow the sample to sit on ice, 1 h; centrifuge at 12,000 g, 15 min

supernatant   pellet

0° C | add slowly with stirring 7.63 g solid $(NH_4)_2SO_4$ / 100 ml; allow the sample to sit on ice, 1 h; centrifuge at 12,000 g, 15 min

supernatant   **pellet**

4° C | dissolve in 100 ml DEAE buffer; dialyze against 4 x 2 liters DEAE buffer, 24 h; centrifuge at 12,000 g, 30 min

supernatant   pellet

4° C | add 10 ml of each PI-cocktail / 10 ml sample; chromatograph on

DEAE-cellulose column

4° C | gradient: 2000 ml (0-0.3 M KCl); dialyze against 2 x 4 liters hydroxylapatide buffer, 18 h; centrifuge at 12,000 g, 30 min;

supernatant   pellet

4° C | chromatograph on

hydroxylapatide column

4° C | put 100 ml of 1 mM EDTA, pH 7.3 in each tube and 100 ml of 1 mM EGTA, pH 7.3, gradient : 300 ml (10-300 mM NaH_2PO_4); dialyze 16 h against 2 liters of HPLC buffer plus 10% sucrose up; filter through a 0.2-mm acrodisc filter; fractionate the sample (12 ml) by

HPLC column

RT | gradient: 40 ml (0-0.3 M KCl)

METAVINCULIN
(~3 mg)

MYOMESIN

A 165 kDa protein with an elongated molecular dimension of 4 x 36 nm, involved in constituating the M-line in muscle sarcomers.

Source:	chicken breast muscle
Equipment:	• meat grinder • centrifuge • vacuum dialysis equipment • DEAE-cellulose (DE-52) column (2.5 x 35 cm) • 5´-AMP-Sepharose 4B (Pharmacia) column (0.5 x 6 cm)
Chemicals:	KCl, EGTA, EDTA, DTT, NaOH, NaCl, $MgCl_2$, $Na_4P_2O_7$, K/PO$_4$, $(NH_4)_2SO_4$ ultrapure **protease inhibitors:** PMSF, pepstatin A
Have ready:	**washing buffer:** 0.1 M KCl, 1 mM EGTA, 5 mM EDTA, 1 mM DTT, pH 7.0 (with NaOH) + 0.1 mM PMSF + 10^{-6} M pepstatin A
	extraction buffer: 0.6 M KCl, 1 mM EDTA, 1 mM $MgCl_2$, 10 mM $Na_4P_2O_7$ (sodium pyrophosphate), 0.3 mM DTT, 0.1 mM potassium phosphate, pH 6.4 + 0.1 mM PMSF + 10^{-6} M pepstatin A
	cold $(NH_4)_2SO_4$: 91% (475 g/l) + 1 mM EDTA, pH 7.2
	buffer A: 0.6 M KCl, 1 mM EDTA, 0.3 mM DTT, 0.05 M potassium phosphate, pH 6.9 + 0.1 mM PMSF, 10^{-6} M pepstatin A
	buffer B: 1 mM EDTA, 0.3 mM DTT, 10 mM potassium phosphate, pH 6.9 + 0.1 mM PMSF, 10^{-6} M pepstatin A
	buffer C: 1 mM EDTA, 0.3 mM DTT, 10 mM potassium phosphate, pH 6.9 + 0.1 mM PMSF, 10^{-6} M pepstatin A
Reference:	Eppenberger, H.M. and Strehler, E. (1982) Methods Enzymol. 85, 143-149

chicken breast muscle (~350 g)

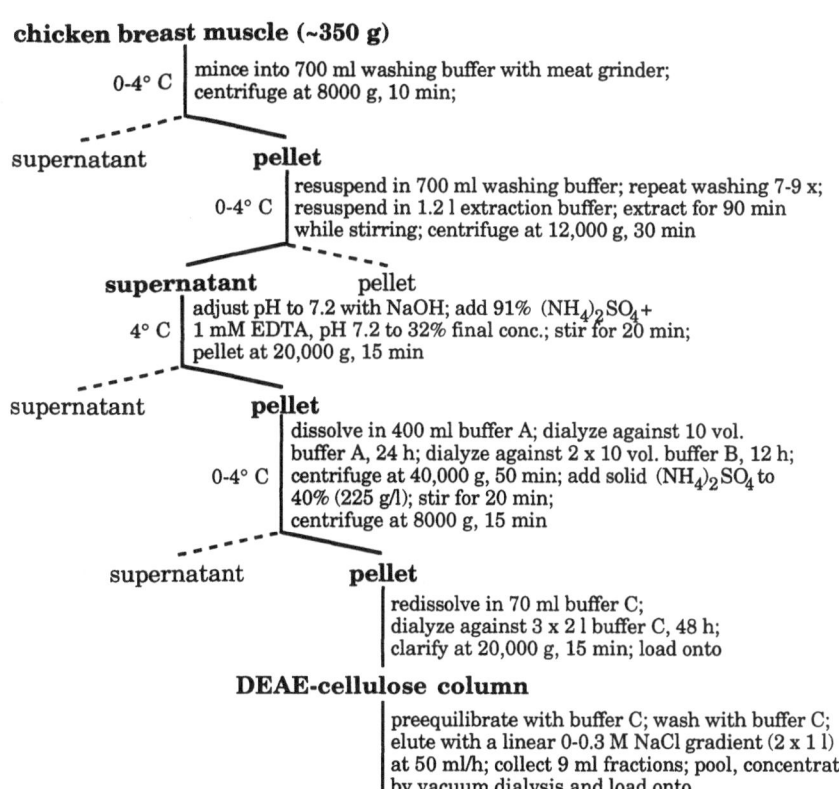

0-4° C | mince into 700 ml washing buffer with meat grinder;
centrifuge at 8000 g, 10 min;

supernatant   **pellet**

0-4° C | resuspend in 700 ml washing buffer; repeat washing 7-9 x;
resuspend in 1.2 l extraction buffer; extract for 90 min
while stirring; centrifuge at 12,000 g, 30 min

supernatant   pellet

4° C | adjust pH to 7.2 with NaOH; add 91% $(NH_4)_2SO_4$ +
1 mM EDTA, pH 7.2 to 32% final conc.; stir for 20 min;
pellet at 20,000 g, 15 min

supernatant   **pellet**

0-4° C | dissolve in 400 ml buffer A; dialyze against 10 vol.
buffer A, 24 h; dialyze against 2 x 10 vol. buffer B, 12 h;
centrifuge at 40,000 g, 50 min; add solid $(NH_4)_2SO_4$ to
40% (225 g/l); stir for 20 min;
centrifuge at 8000 g, 15 min

supernatant   **pellet**

redissolve in 70 ml buffer C;
dialyze against 3 x 2 l buffer C, 48 h;
clarify at 20,000 g, 15 min; load onto

DEAE-cellulose column

preequilibrate with buffer C; wash with buffer C;
elute with a linear 0-0.3 M NaCl gradient (2 x 1 l)
at 50 ml/h; collect 9 ml fractions; pool, concentrate
by vacuum dialysis and load onto

5´-AMP-Sepharose 4B column

equilibrate in buffer C;
recover myomesin in break
through fractions

MYOMESIN
(40-50 mg total, 0.25 mg/ml)

NUCLEAR ACTIN BINDING (NAB) -PROTEIN

A nuclear actin binding protein consisting of two 34 kDa polypeptides. It binds to DNA and actin but has no cross-linking, capping or severing activity. No ATP-ase activity though antigenically related to myosin I.

Source: Acanthamoeba c.

Equipment:
- Potter-Elvijhem homogenizer
- centrifuge
- JA-10, J6, JA-20 rotor
- Dounce (glass) homogenizer
- hydroxylapatide column (BioRad Labs, Richmond CA) (50 ml)
- sephacryl-S300 column (1.5 x 50 cm)
- heparin-agarose column (Sigma) (10 ml)
- SDS-PAGE

Chemicals: imidazole-HCl, sucrose, NaCl, EGTA, ATP, DTT, Tris-HCl, $MgCl_2$, KCl

protease inhibitors: PMSF, benzamidine

Have ready: **extraction buffer:** 10% sucrose, 20 mM imidazole-HCl, 1 mM EGTA, 1 mM ATP, 1 mM DTT, 0.1 mM benzamidine, 0.1 mM PMSF, pH 7.5

STMDK-buffer: 0.75 M sucrose, 40 mM Tris-HCl, 4 mM $MgCl_2$, 50 mM KCl, 1 mM DTT, pH 7.1

HSE-buffer: 0.7 M NaCl, 30 mM KCl, 5 mM $MgCl_2$, 50 mM Tris-HCl, 0.5 mM DTT, 0.1 mM benzamidine, pH 8.0

TD-buffer: 2 mM Tris-HCl, 0.5 mM DTT, pH 8.0

Reference: Rimm, D.L. and Pollard, T.D. (1989) J. Cell Biol. 109, 585-591

acanthamoeba cells (~900 g)

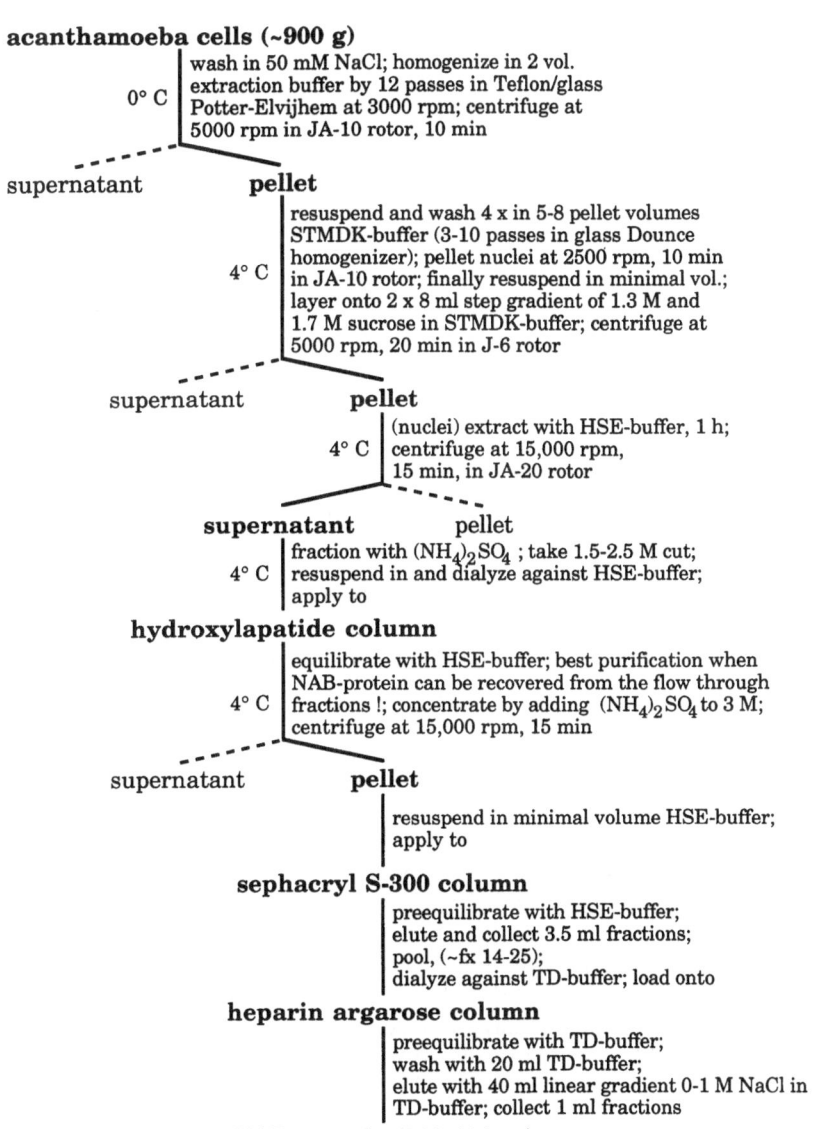

0° C | wash in 50 mM NaCl; homogenize in 2 vol. extraction buffer by 12 passes in Teflon/glass Potter-Elvijhem at 3000 rpm; centrifuge at 5000 rpm in JA-10 rotor, 10 min

- - - - supernatant

pellet

4° C | resuspend and wash 4 x in 5-8 pellet volumes STMDK-buffer (3-10 passes in glass Dounce homogenizer); pellet nuclei at 2500 rpm, 10 min in JA-10 rotor; finally resuspend in minimal vol.; layer onto 2 x 8 ml step gradient of 1.3 M and 1.7 M sucrose in STMDK-buffer; centrifuge at 5000 rpm, 20 min in J-6 rotor

- - - - supernatant

pellet

4° C | (nuclei) extract with HSE-buffer, 1 h; centrifuge at 15,000 rpm, 15 min, in JA-20 rotor

- - - -

supernatant   pellet

4° C | fraction with $(NH_4)_2SO_4$; take 1.5-2.5 M cut; resuspend in and dialyze against HSE-buffer; apply to

hydroxylapatide column

4° C | equilibrate with HSE-buffer; best purification when NAB-protein can be recovered from the flow through fractions !; concentrate by adding $(NH_4)_2SO_4$ to 3 M; centrifuge at 15,000 rpm, 15 min

- - - - supernatant

pellet

resuspend in minimal volume HSE-buffer; apply to

sephacryl S-300 column

preequilibrate with HSE-buffer; elute and collect 3.5 ml fractions; pool, (~fx 14-25); dialyze against TD-buffer; load onto

heparin argarose column

preequilibrate with TD-buffer; wash with 20 ml TD-buffer; elute with 40 ml linear gradient 0-1 M NaCl in TD-buffer; collect 1 ml fractions

NAB-protein (200-600 μg)
(store in TD-buffer at 4° C)

NEBULIN

A family of 650-850 kDa polymer presumably acting as length regulator for thin filaments within skeletal muscle sarcomers.

Source: striated muscle

Equipment:
- Waring blender
- centrifuge, GSA rotor
- high porosity SDS-gel system (see below)
- water bath
- Bio Gel A-150 M column (2.6 x 80 cm)
- Amicon concentration set, PM 30 filter

Chemicals: KCl, $MgCl_2$, EGTA, EDTA, Tris-maleate,
DTT, Triton X-100, SDS, sodium acetate, Tris-HCl,
$Na_4P_2O_7$, glycerol, pyronin, Coomassie blue R,
isopropanol, acetic acid, Tris-acetate
protease inhibitors: PMSF

Have ready: **PRB-buffer (pyrophosphate relaxing buffer):**
0.1 M KCl, 2 mM $MgCl_2$, 2 mM EGTA, 10 M Tris-
maleate, 0.5 mM DTT, 0.1 mM PMSF, 2 mM
$Na_4P_2O_7$, pH 6.8

low salt buffer: PRB-buffer without pyrophosphate

Triton-buffer: low salt buffer + 0.5% Triton X-100

washing buffer: 5 mM Tris-HCl, pH 8.0
NaCl: 4 mM in wash buffer
SDS 3% in wash buffer

gel filtration column buffer: 40 mM Tris-HCl,
20 mM sodium acetate, 2 mM EDTA, 0.1 mM DTT,
0.1% SDS, pH 7.4

High porosity gel system:
polyacrylamide gel: 0.6 x 10 cm tube gels of
3.2% acrylamide (acrylamide/bisacrylamide:
50/1), 0.1% TEMED, 0.06% ammonium
persulfate
gel and electrophoresis buffer: 40 mM Tris-
acetate, 20 mM sodium acetate, 2 mM EDTA,
0.1% SDS, pH 7.4
3x SDS sample buffer: 30 mM Tris-HCl, 3 mM
EDTA, 3% SDS, 120 mM DTT, 30% glycerol, 3 µg
pyronin Y/ml, pH 8.0
staining solution: 0.1% Coomassie Blue R, 25%
isopropanol, 10% acetic acid
destaining solution: 5% isopropanol, 10% acetic acid

Reference: Wang, K. (1982) Methods Enzymol. 85, 264-274
Gen-Bank: X58122 (18 kB human nebulin)

**see salt fractionation
from titin preparation**

supernatant   pellet

20° C │ to 32 ml supernatant add 0.5 ml 4 M NaCl to make
      │ 0.7 M NaCl; after 10 min centrifuge at 45,000 rpm,
      │ 30 min, in Ti 50 rotor at 20°C

supernatant   pellet

│ concentrate by ultrafiltration with Amicon PM 30 filter;
│ apply to

Bio Gel A-150 column

│ equilibrated with gel filtration column buffer;
│ collect 10 ml fractions;
│ nebulin elutes in the ascending second peak

NEBULIN (4-5 mg)

PARAMYOSIN

A 210 kDa two chain α-helical coiled-wil protein of 120 nm in length forming the core of invertebrate thick myosin filaments.

Source:	Merceneria (horse shoe crab)
Equipment:	• Potter homogenizer • cheese cloth • centrifuge
Chemicals:	KI, Tris-HCl, DTT, $(NH_4)_2SO_4$, KH_2PO_4, K_2HPO_4, 0.1 M HCl, KOH, KCl
Have ready:	**extraction solution:** 0.3 M KI, 20 mM Tris, 0.1 mM DTT, pH 8.3
	1 M KH_2PO_4, cold
	ice-cold, deionized H_2O
	cold saturated $(NH_4)_2SO_4$
	homogenization solution I: 0.3 M KI, 60 mM potassium phosphate buffer, 0.1 mM DTT, pH 8.3
	homogenization solution II: 0.6 M KCl, 60 mM potassium phosphate buffer, 0.1 mM DTT, pH 7.4
	homogenization solution III: 0.6 M KCl, 10 mM potassium phosphate buffer, 0.1 mM DTT, pH 7.3
	phosphate buffer: 10 mM potassium phosphate buffer, 0.1 mM DTT, pH 6.0
	solubilization buffer: 0.6 M KCl, 10 mM potassium phosphate buffer, 0.1 mM DTT, pH 7.3
Reference:	Levine, R.J.C. et al. (1982) Methods Enzymol. 85, 149-160

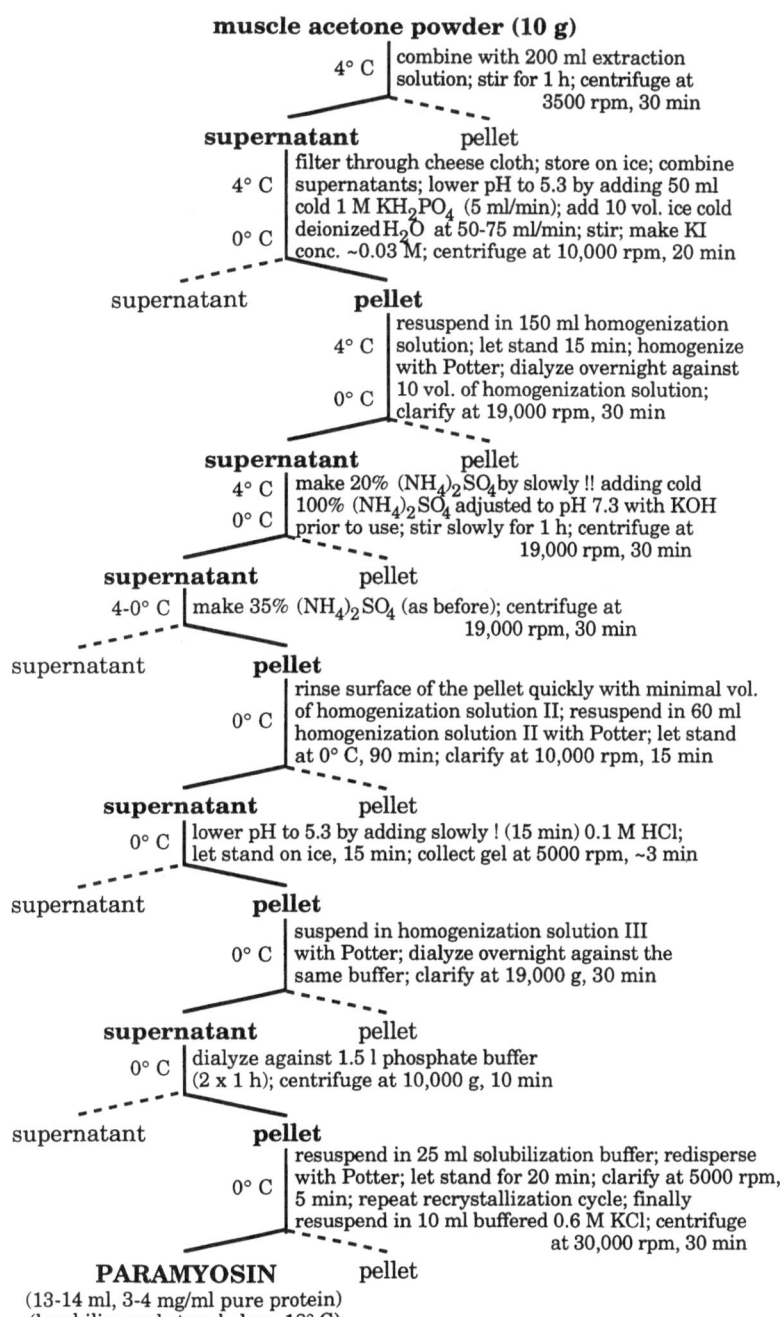

muscle acetone powder (10 g)

4° C | combine with 200 ml extraction solution; stir for 1 h; centrifuge at 3500 rpm, 30 min

supernatant   pellet

4° C | filter through cheese cloth; store on ice; combine supernatants; lower pH to 5.3 by adding 50 ml cold 1 M KH_2PO_4 (5 ml/min); add 10 vol. ice cold
0° C | deionized H_2O at 50-75 ml/min; stir; make KI conc. ~0.03 M; centrifuge at 10,000 rpm, 20 min

supernatant   **pellet**

4° C | resuspend in 150 ml homogenization solution; let stand 15 min; homogenize with Potter; dialyze overnight against
0° C | 10 vol. of homogenization solution; clarify at 19,000 rpm, 30 min

supernatant   pellet

4° C | make 20% $(NH_4)_2SO_4$ by slowly !! adding cold 100% $(NH_4)_2SO_4$ adjusted to pH 7.3 with KOH
0° C | prior to use; stir slowly for 1 h; centrifuge at 19,000 rpm, 30 min

supernatant   pellet

4-0° C | make 35% $(NH_4)_2SO_4$ (as before); centrifuge at 19,000 rpm, 30 min

supernatant   **pellet**

0° C | rinse surface of the pellet quickly with minimal vol. of homogenization solution II; resuspend in 60 ml homogenization solution II with Potter; let stand at 0° C, 90 min; clarify at 10,000 rpm, 15 min

supernatant   pellet

0° C | lower pH to 5.3 by adding slowly ! (15 min) 0.1 M HCl; let stand on ice, 15 min; collect gel at 5000 rpm, ~3 min

supernatant   **pellet**

0° C | suspend in homogenization solution III with Potter; dialyze overnight against the same buffer; clarify at 19,000 g, 30 min

supernatant   pellet

0° C | dialyze against 1.5 l phosphate buffer (2 x 1 h); centrifuge at 10,000 g, 10 min

supernatant   **pellet**

0° C | resuspend in 25 ml solubilization buffer; redisperse with Potter; let stand for 20 min; clarify at 5000 rpm, 5 min; repeat recrystallization cycle; finally resuspend in 10 ml buffered 0.6 M KCl; centrifuge at 30,000 rpm, 30 min

PARAMYOSIN   pellet

(13-14 ml, 3-4 mg/ml pure protein)
(lyophilize and store below -18° C)

PAXILLIN

A tyrosine phosphorylated 68 kDa vinculin binding protein present in focal cell adhesions and dense plaques of smooth muscle. Supposed to be involved in regulating protein assembly at cytoskeleton-membrane interactions.

Source: chicken gizzard

Equipment:
- Waring blender
- centrifuge
- cheese cloth
- DEAE-cellulose column (DE-52) (100 ml)
- hydroxylapatide column (25 ml)
- paxillin antibody-Affi gel-10 affinity column

Chemicals: Tris-HCl, EGTA, 2-mercaptoethanol, Tris-acetate, EDTA, NaCl, K/PO$_4$, Na/PO$_4$, glycine-HCl

protease inhibitors: PMSF, leupeptin

Have ready:

μM

homogenization buffer: 10 mM Tris-HCl, 2 mM EGTA, 0.1% 2-mercaptoethanol, 0.5 mM PMSF, 5 leupeptin, pH 8.0

buffer B: 20 mM Tris-acetate, 20 mM NaCl, 0.1 mM EDTA, 0.1% 2-mercaptoethanol, pH 7.6 + 0.5 mM PMSF, 5 μM leupeptin

Reference: Turner, C.E. et al. (1990) J. Cell Biol. 111, 1059-1068

chicken gizzard (~150 g)

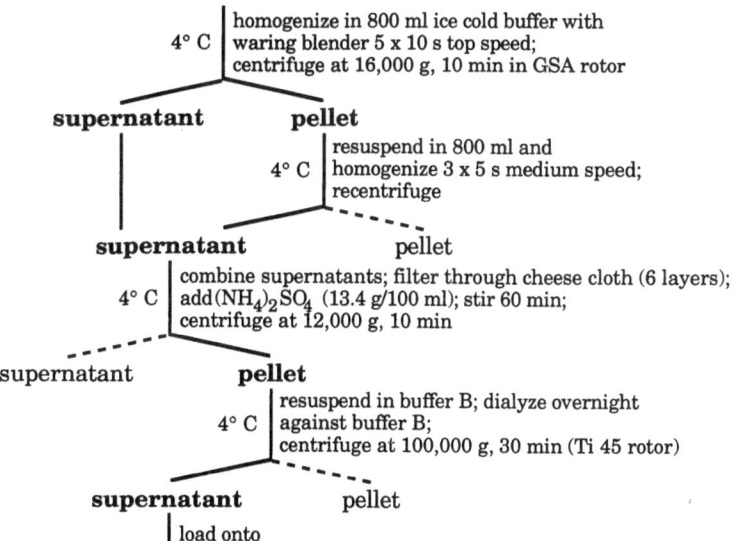

4° C | homogenize in 800 ml ice cold buffer with
waring blender 5 x 10 s top speed;
centrifuge at 16,000 g, 10 min in GSA rotor

supernatant   **pellet**

4° C | resuspend in 800 ml and
homogenize 3 x 5 s medium speed;
recentrifuge

supernatant   pellet

4° C | combine supernatants; filter through cheese cloth (6 layers);
add $(NH_4)_2SO_4$ (13.4 g/100 ml); stir 60 min;
centrifuge at 12,000 g, 10 min

supernatant   **pellet**

4° C | resuspend in buffer B; dialyze overnight
against buffer B;
centrifuge at 100,000 g, 30 min (Ti 45 rotor)

supernatant   pellet

| load onto

DEAE-cellulose column

preequilibrate with buffer B; wash with 300 ml bufferB;
elute with 650 ml total (buffer B - buffer B + 325 mM NaCl);
pool paxillin rich fractions and load directly on

hydroxylapatide column

preequilibrated in buffer B; elute with 120 ml linear gradient
(15-240 mM potassium phosphate, pH 7.5 + 0.1%
2-mercaptoethanol); pool fractions around ~30 mM phosphate
and dialyze against Tris buffered saline (TBS), pH 7.6;
load onto

**paxillin antibody-Affi gel-10
affinity column**

wash with PBS, then with 10 mM sodium phosphate, pH 6.8;
elute with 100 mM glycine-HCl, pH 2.5 into tubes containing
Tris-HCl, pH 10.0 + 0.2% 2-mercaptoethanol

PAXILLIN

PONTICULIN

An integral, monomeric membrane protein of 17 kDa which binds to
F-actin and nucleates filament polymerization.

Source:
Dictyostelium discoideum plasma membranes

Equipment:
- F-Actin affinity column
  (anti fluorescein IgG-sephacryl S-1000)
- F-Actin affinity matrix
- centrifuge
- 50 ml plastic conical tube
- 12 ml and 30 ml plastic syringe fitted with a
  plug of porous polyethylene
- Amicon: centriprep-10, and centricon-10
  microconcentrators
- spectrapor-1 dialysis tubing M_r 6000-8000
- electrophoretic gels
- preparative SDS polyacrylamide gels

Chemicals:
KCl, $MgCl_2$, PIPES, NaN_3, DTT, ATP, fluorescein-
labeled actin, unlabeled actin, phalloidin, DMSO,
Trizma-Base, acetic acid, octylglucoside (OG), NaCl,
EGTA, BSA
column materials: affinity column, anti-fluorescein
IgG coupled to sephacryl S-1000

Have Ready:
F-actin affinity column

column buffer: 50 mM KCl, 1 mM $MgCl_2$, 20 mM
Tris-acetat, pH 7.0 at 20° C (1 liter)

polymerization buffer (5x): 500 mM KCl, 10 mM
$MgCl_2$, 100 mM PIPES, 0.02% NaN_3, pH 7.0 (2.7 ml)

30%, 6%, 3%, octylglucoside in column buffer

high salt buffer: 2 M NaCl, 2 mM EGTA, 1%
octylglucoside in column buffer (15 ml)

blocking solution: 4% (w/v) BSA, 0.5 mM
phalloidin in column buffer (10 ml)

Reference:
Wuesthube, L., Speicher, D., Shariff, A. and
Luna, E. (1991) Methods Enzymol. 196, 47-65

Dictyostelium discoideum plasma membranes

4-6° C | solubilize in an equal volume of 6% OG in column buffer; dilute with 3% OG in column buffer to a final protein concentration of 0.2 mg/ml, incubate 10-20 min; centrifuge at 100,000 g, 1 h

supernatant   pellet

RT | preelute preparative F-actin affinity column with 5 ml high salt buffer; wash with 10 ml column buffer; block nonspecific protein-binding sites with 12 ml blocking solution

RT | incubate the column material with 10 mg / 50 ml solubilized, clarified plasma membrane; collect the run-through fraction and reincubate it after first elution; wash the column with 10 ml 1% OG in column buffer; elute salt sensitive actin binding proteins with 5 ml high salt buffer

salt sensitive actin binding proteins

4° C | concentrate approxymately 10-fold in a centriprep-10 unit to about 1ml

filtrate   **retentate**

4° C | rinse with 1 ml 1% OG in column buffer; concentrate rinse and retentate in a centricon-10 microconcentrator to 0.5-1 ml by centrifugation at 4300 g, 2 h

ponticulin sample

4° C | dialyze in a spectrapor-1 dialysis bag (with a retention range Mt 6000 to 8000, allowing sample room for ca. twofold expansion of the sample volume) against 50 ml 1% OG in column buffer for 22-48 h

ponticulin sample

4° C | purify the sample on preparative SDS-polyacrylamide gels; dialyze and add NaN_3 to a final concentration of 0.02%

PONTICULIN

PROFILIN

A small, globular 12,000-14,000 MW protein of 124-140 amino acids which binds actin and PiP-2. Possibly regulating actin assembly at membranes.

Source:	platelets
Equipment:	• centrifuge • Ultrasonics W 185 F sonicator • poly(L-proline) (PLP) sepharose column (10 ml) • electrophoretic gels
Chemicals:	Trizma-Base, KCl, DTT, ATP, urea (up), NaCl, $NaHCO_3$, glycine, nonionic detergents **protease inhibitors:** benzamidin, PMSF, aprotinin **column materials:** PLP M_r 40,000, swollen hydrated beads
Have ready:	**PLP column argarose** **extraction buffer I:** 20 mM Tris-HCl, 150 mM KCl, 1 mM PMSF, 0.2 mM ATP, 0.2 mM DTT, pH 7.4 **elution buffer II (actin):** buffer I plus 2 M urea **elution buffer III:** buffer I plus 4 M urea **elution buffer IV (profilin):** buffer I plus 7 M urea
Reference: A.C.	Lindberg, U. Schutt, C.E., Hellsten, E., Tjader, and Hult, T. (1988) BBA 967, 391-400 Janmey, P.A. (1991) Methods Enzymol. 196, 92-99

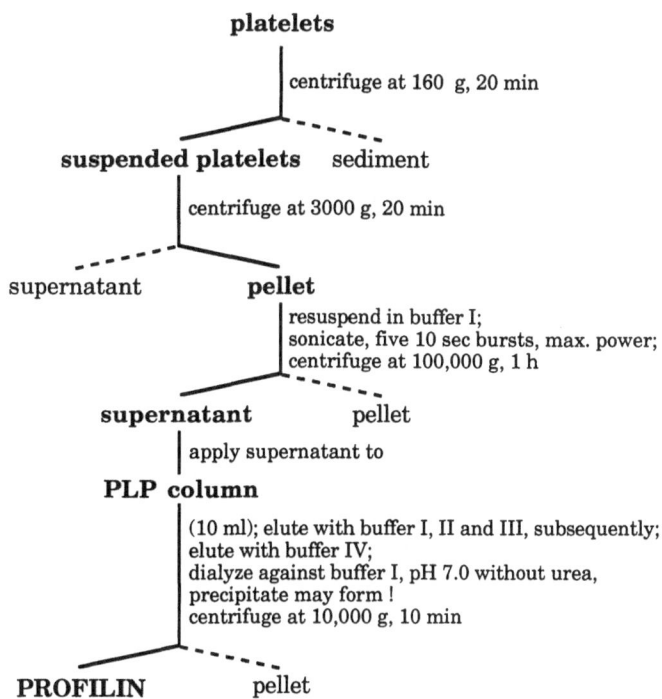

platelets

centrifuge at 160 g, 20 min

suspended platelets   sediment

centrifuge at 3000 g, 20 min

supernatant   **pellet**

resuspend in buffer I;
sonicate, five 10 sec bursts, max. power;
centrifuge at 100,000 g, 1 h

supernatant   pellet

apply supernatant to

PLP column

(10 ml); elute with buffer I, II and III, subsequently;
elute with buffer IV;
dialyze against buffer I, pH 7.0 without urea,
precipitate may form !
centrifuge at 10,000 g, 10 min

PROFILIN   pellet

PROTEIN 4.1

A 82 kDa actin, spectrin and lipid binding protein involved in linking the spectrin-actin lattice to membranes. Substrate for cAMP and protein kinase C catalyzed phosphorylation.

Source: human erythrocytes

Equipment:
- centrifuge
- water bath
- DEAE-cellulose column (1.6 x 30 cm)

Chemicals: Na/PO$_4$, EDTA, KCl, DTT, NaCl, NaN$_3$
protease inhibitors: PMSF, DFP

Have ready: **lysis buffer:** 5 mM Na/PO$_4$, 1 mM EDTA, 0.5 mM PMSF, pH 8.0

PBS: 5 mM Na/PO$_4$, 1 mM EDTA, 155 mM NaCl, pH 7.6 + 0.4 mM (0.0075%) DFP

washing solution: 0.3 mM Na/PO$_4$, 0.2 mM EDTA, pH 7.6

1 M borate buffer, pH 8.5

DEAE-column buffer: 5 mM Na/PO$_4$, 1 mM EDTA, 20 mM KCl, 0.5 mM DTT, pH 7.6

dialyzing buffer: 5 mM Na/PO$_4$, 1 mM EDTA, 130 mM KCl, 20 mM NaCl, 0.2 mM DTT, 2 mM NaN$_3$, pH 7.6

Reference: Tyler, J.M. et al (1980) J. Biol. Chem. 255, 7034-7039

50 ml packed human erythrocytes

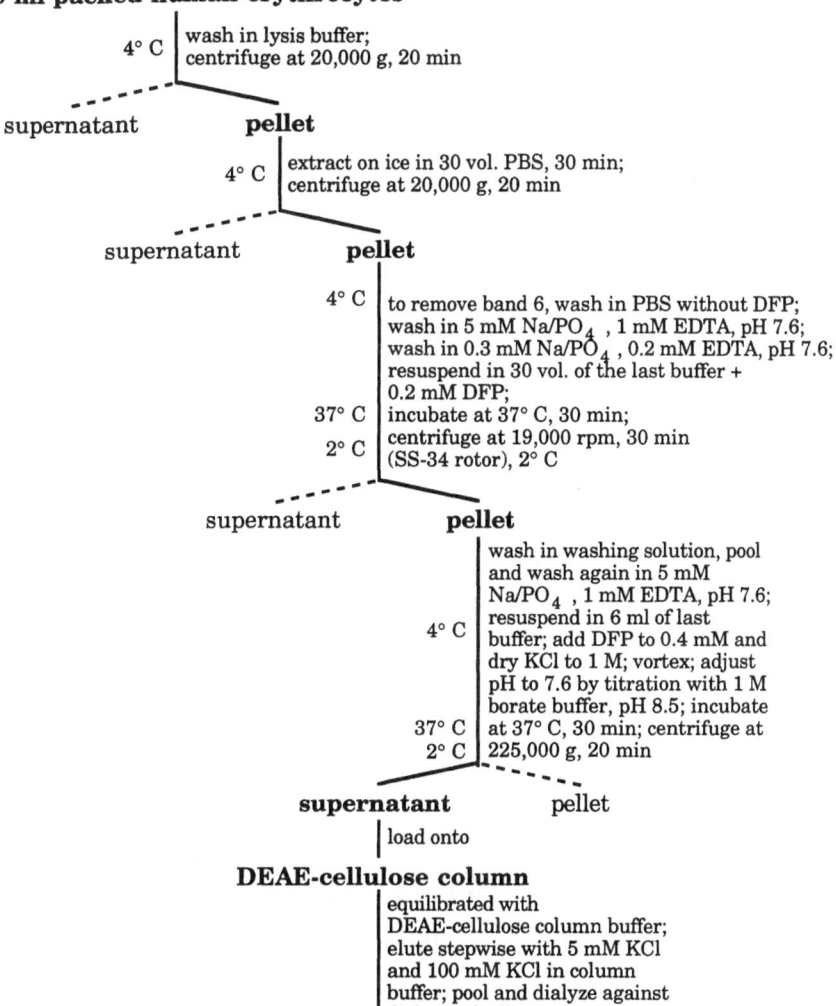

supernatant

PROTEIN 4.1

RADIXIN

A 82 kDa barbed-end actin filament capping protein concentrated in cell adherens junctions and the cleavage furrow. Presumably an end-on actin-filament/membrane linker protein.

Source: chicken gizzard

Equipment:
- Waring blender
- centrifuge
- DEAE cellulose (DE 52) column (1.5 x 5 cm)
- DNA-se I-actin column

Chemicals: Tris-HCl, EGTA, hydrochloric acid, MgCl$_2$, HEPES, DTT, ATP, KCl, Affi-gel 10, DNA-se I
protease inhibitors: PMSF, leupeptin, TAME (p-toluenesulfonyl-L-arginine methyl ester hydrochloride)

Have ready: **dissection solution:** distilled H$_2$O, 1 mM PMSF, 1 µg/ml leupeptin

solution A: 2 mM Tris-HCl, 1 mM EGTA, 0.5 mM PMSF, 1 µg/ml leupeptin, pH 9.0

solution B: 10 mM HEPES, 1 mM EGTA, 0.5 mM PMSF, 1 µg/ml leupeptin, 0.1 mM DTT, pH 7.5

solution C: 2 mM HEPES, 0.2 mM ATP, 0.2 mM DTT, 1 mM TAME, 1 mM PMSF, 1 µg/ml leupeptin, pH 7.5

solution D: equal volumes of solution B and C

1.3 M MgCl$_2$ solution

DNA-se I-actin column: mix 5 mg gel-filtered G-actin with DNA-se I coupled Affi-gel 10

Reference: Sato, N., Yonemura, S., Obinato, T., Tsukita, S. and S. Tsukita (1991) J. Cell Biol. 113, 321-330

116

chicken gizzard smooth muscle

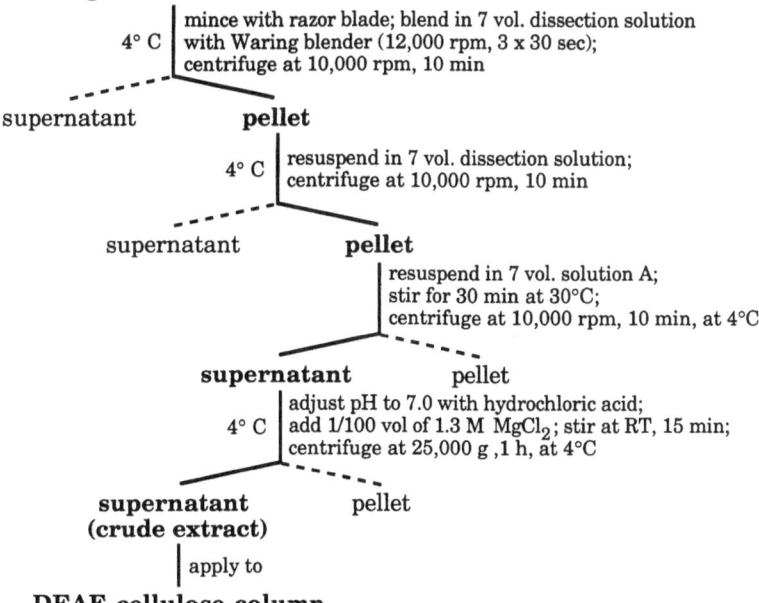

4° C | mince with razor blade; blend in 7 vol. dissection solution with Waring blender (12,000 rpm, 3 x 30 sec); centrifuge at 10,000 rpm, 10 min

supernatant   **pellet**

4° C | resuspend in 7 vol. dissection solution; centrifuge at 10,000 rpm, 10 min

supernatant   **pellet**

resuspend in 7 vol. solution A; stir for 30 min at 30°C; centrifuge at 10,000 rpm, 10 min, at 4°C

supernatant   pellet

4° C | adjust pH to 7.0 with hydrochloric acid; add 1/100 vol of 1.3 M MgCl$_2$; stir at RT, 15 min; centrifuge at 25,000 g ,1 h, at 4°C

supernatant   pellet
(crude extract)

apply to

DEAE cellulose column

equilibrated with solution B; wash thoroughly; elute with solution B + 60 mM KCl; dilute eluate with equal volume solution C; load onto

DNA-se I-actin column

equilibrated with solution D, solution D+400 mM KCl and solution D sequentially; after loading wash with solution D; elute with solution D+160 mM KCl

RADIXIN
(3-5 µg/30 g chicken gizzard)

SCINDERIN

A 80 kDa Ca⁺⁺-activated F-actin severing protein from chromaffin cells with three isoforms. Different from gelsolin.

Source:	bovine adrenal medulla
Equipment:	• Potter Elvehjem homogenizer • centrifuge • DEAE-cellulose column (2.5 x 35 cm) • sephadex-G-100 column (2.5 x 90 cm) • DNA-se-I-actin-sepharose 4B affinity column (0.9 x 10 cm) • DEAE-5 PW column (Millipore) (0.75 x 7.5 cm) connected to water HPLC system
Chemicals:	sucrose, imidazole, DTT, NaN$_3$, EGTA, ATP, KCl, Tris-HCl, DNA-se I **protease inhibitors:** DFP, PMSF, leupeptin, soyabean trypsin inhibitor
Have ready:	**extraction solution:** 0.3 M sucrose, 20 mM imidazole, 5 mM DTT, 1 mM PMSF, 0.75 mM NaN$_3$, 0.1 mM EGTA, 1 mM ATP, 300 mM KCl, 10 µg/ml soyabean trypsin inhibitor, 1 µg/ml leupeptin, 0.1 mM DFP, pH 7.5 **buffer A:** 20 mM Tris-HCl, 0.1 mM DTT, 0.1 mM EGTA, pH 7.5 **buffer B:** 1 M KCl in buffer A
Reference:	Rodriguez Del Castillo, A. et al. (1990) EMBO J. 9, 43-52

adrenal medullae (70-100 g, 25-30 glands)

4° C | homogenize in 4 ml/g extraction solution;
centrifuge at 1000 g, 10 min

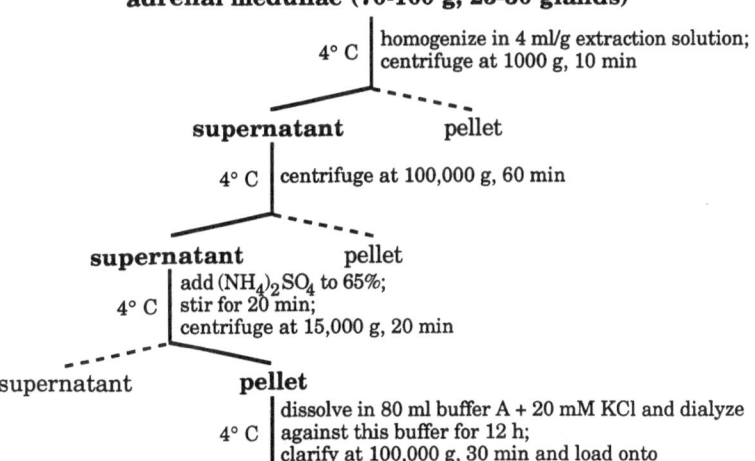

supernatant   pellet

4° C | centrifuge at 100,000 g, 60 min

supernatant   pellet

4° C | add (NH$_4$)$_2$SO$_4$ to 65%;
stir for 20 min;
centrifuge at 15,000 g, 20 min

supernatant   **pellet**

4° C | dissolve in 80 ml buffer A + 20 mM KCl and dialyze
against this buffer for 12 h;
clarify at 100,000 g, 30 min and load onto

DEAE-cellulose column

4° C | equilibrated in buffer A + 20 mM KCl;
elute at 20 ml/h with linear gradient 0.02-1 M KCl in
buffer A (540 ml);
collect 6 ml fractions; pool, concentrate and apply to

sephadex-G-100-column

4° C | equilibrated with buffer A + 100 mM KCl + 2 mM
CaCl$_2$; wash with buffer A + 0.5 M KCl; elute with
buffer A + 20 mM KCl + 10 mM EDTA;
pool, concentrate and apply to

DEAE-5 PW HPLC column

4° C | elute at 0.8 ml/min with buffer A + 20 mM KCl, 10 min,
then with linear gradient 0-10% buffer B, 10-50% buffer B
and 50-100% buffer B for 10 min;
collect 0.8 ml fractions

SCINDERIN

SEVERIN

A 40 kDa monomeric protein which severs, caps and nucleates actin filaments in a Ca^{++}-dependent manner.

Source: Dictyostelium discoideum

Equipment:
- sonicator
- centrifuge
- DEAE cellulose (DE-52) column (4 x 30 cm)
- hydroxylapatide column (2.5 x 25 cm)
- sephacryl S-200 column (1.5 x 80 cm)

Chemicals: triethanolamine, EGTA, $Na_4P_2O_7$, sucrose, Trasylol (Boehringer), PMSF, DTT, KCl, K/PO$_4$

Have ready: **lysis buffer:** 10 mM triethanolamine, 60 mM sodiumpyrophosphate, 30% sucrose, 0.01 mg/ml Trasylol, 3 mM EGTA, 0.5 mM PMSF

dialyzing buffer: 2 mM triethanolamine, 0.2 mM DTT, pH 7.5

hydroxylapatide column buffer: 10 mM K/PO$_4$, 0.2 mM DTT, pH 7.6

sephacryl S-200 column buffer: 20 mM triehanolamine, 0.2 mM DTT, 50 mM KCl, pH 7.5

Reference: Brown, S.S. (1986) Methods Enzymol. 134, 9-13
Gen-Bank: J03515

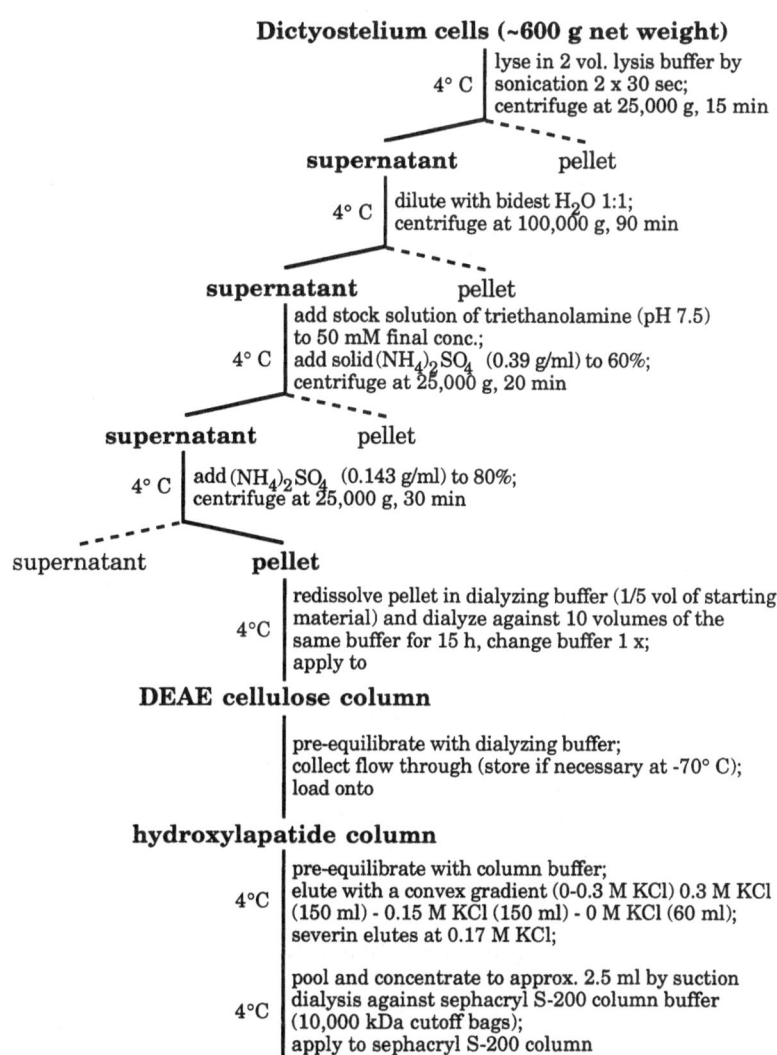

Dictyostelium cells (~600 g net weight)

4° C | lyse in 2 vol. lysis buffer by sonication 2 x 30 sec; centrifuge at 25,000 g, 15 min

supernatant   pellet

4° C | dilute with bidest H_2O 1:1; centrifuge at 100,000 g, 90 min

supernatant   pellet

4° C | add stock solution of triethanolamine (pH 7.5) to 50 mM final conc.; add solid $(NH_4)_2SO_4$ (0.39 g/ml) to 60%; centrifuge at 25,000 g, 20 min

supernatant   pellet

4° C | add $(NH_4)_2SO_4$ (0.143 g/ml) to 80%; centrifuge at 25,000 g, 30 min

supernatant   **pellet**

4°C | redissolve pellet in dialyzing buffer (1/5 vol of starting material) and dialyze against 10 volumes of the same buffer for 15 h, change buffer 1 x; apply to

DEAE cellulose column

| pre-equilibrate with dialyzing buffer; collect flow through (store if necessary at -70° C); load onto

hydroxylapatide column

4°C | pre-equilibrate with column buffer; elute with a convex gradient (0-0.3 M KCl) 0.3 M KCl (150 ml) - 0.15 M KCl (150 ml) - 0 M KCl (60 ml); severin elutes at 0.17 M KCl;

4°C | pool and concentrate to approx. 2.5 ml by suction dialysis against sephacryl S-200 column buffer (10,000 kDa cutoff bags); apply to sephacryl S-200 column

SEVERIN (~6 mg)

SPECTRIN

> **Non erythroid spectrin forms heterodimers of α (284 kDa) and β (246 kDa) subunits. They form antiparallel α-helical rods with approx. 106 residue repeats. Part of the submembrane skeleton.**

Source: brain from pigs or calves

Equipment:
- Polytron-homogenizer (large head)
- centrifuge
- syringe without needle 50 cm^3
- sephacryl S-500 column (5 x 85 cm)
- DE-53 cellulose column (1.6 x 15 cm)
- hydroxylapatite column (0.8 x 12 cm)
- electrophoretic gels
- chemical hood

Chemicals: sucrose, EGTA, NaH_2PO_4, EDTA, DTT, NaCl, KCl, Tween 20, Triton X-100, NaBr, NaN_3, DMSO, $Na_4P_2O_7$

protease inhibitors: pepstatin, leupeptin, PMSF, DFP

Have ready: **homogenization buffer:** 0.32 M sucrose, 2 mM EGTA

protease inhibitors: 5 mg/ml leupeptin, 5 mg/ml pepstatin dissolved first in 0.5 ml DMSO, 200 mg/ml PMSF dissolved first at 200 mg/ml in DMSO, DFP 0.02% (v/v), pH 7.5

demyelination buffer: 1 M sucrose, 2 mM EGTA, 100 mg/ml PMSF, pH 7.5

washing buffer I: 10 mM NaH_2PO_4, 1 mM EGTA, 100 mg/ml PMSF, pH 7.5

washing buffer II: 2 mM NaH_2PO_4, 0.2 mM EGTA, 0.4 mM DTT, 20 mg/ml PMSF, pH 7.5

extraction buffer (low ionic strength): 0.25 mM EDTA, 0.5 DTT, 20 mg/ml PMSF, pH 7.5

gel filtration buffer: 1 M NaBr, 10 mM NaH_2PO_4, 1 mM EGTA, 15 mM Na-pyrophosphat, 1 mM NaN_3, 0.4 mM DTT, pH 8.2

DE-53-column buffer: 10 mM NaH_2PO_4, 0.2 mM EDTA, 1 mM NaN_3, 0.4 mM DTT, pH 7.5

Reference: Davis, J. and Bennett, V. (1983) J. Biol. Chem. 258, 7757-7766
Gen-Bank: X14519 (chicken α-spectrin), J05243 (human brain α-spectrin)

**brain from pigs or
calves (300-400 g)**

4° C | chill the brain in homogenisation buffer without protease inhibitors;
dissect the brain; cut the tissue into 1 cm pieces; wash in sucrose;
freeze the brain tissue with liquid nitrogen

frozen brain

4° C | thawe frozen brain in 1.5 liters homogenization buffer with protease
inhibitors, except PMSF and DFP; pellet the suspension at 900 g, 10 min

supernatant   **pellet**

4° C | centrifuge at 30,000 g, 30 min

supernatant   **pellet**

4° C | resuspend in 1.5 liters demyelination buffer;
centrifuge at 30,000 g, 45 min

membranes   supernatant

4° C | resuspend in 1.5 liters washing buffer I;
centrifuge at 30,000 g, 25 min

supernatant   **membranes**

37° C | wash with 1.5 liters washing buffer II;
resuspend in 1.2 liters low ionic buffer;
incubate 60 min at 37° C;
centrifuge at 30,000 g, 40 min

supernatant   pellet

4° C | adjust to 10 mM NaH_2PO_4 pH 7.5;
centrifuge at 30,000 rpm, 60 min

supernatant   pellet

4° C | add 31.3 g $(NH_4)_2SO_4$ /100 ml; chill at -10° C,
30 min; centrifuge at 30,000 g, 10 min

supernatant   **pellet**

4° C | dissolve in 50 ml 10% sucrose gel filtration buffer;
centrifuge at 82,000 g, 15 h

supernatant   pellet

4° C | chromatograph on

sephacryl S 500 column

4° C | add 31.3 g $(NH_4)_2SO_4$ /100 ml; chill at -10° C, 2 h;
centrifuge at 30,000 g, 30 min

supernatant   **pellet**

4° C | redissolve in 15 ml gel filtration buffer;
centrifuge at 50,000 rpm, 60 min

supernatant   pellet

| dialyze against DE-53 buffer; chromatograph on

DE-53 column

| gradient: 100 ml (0.15-0.4 M NaCl)

SPECTRIN

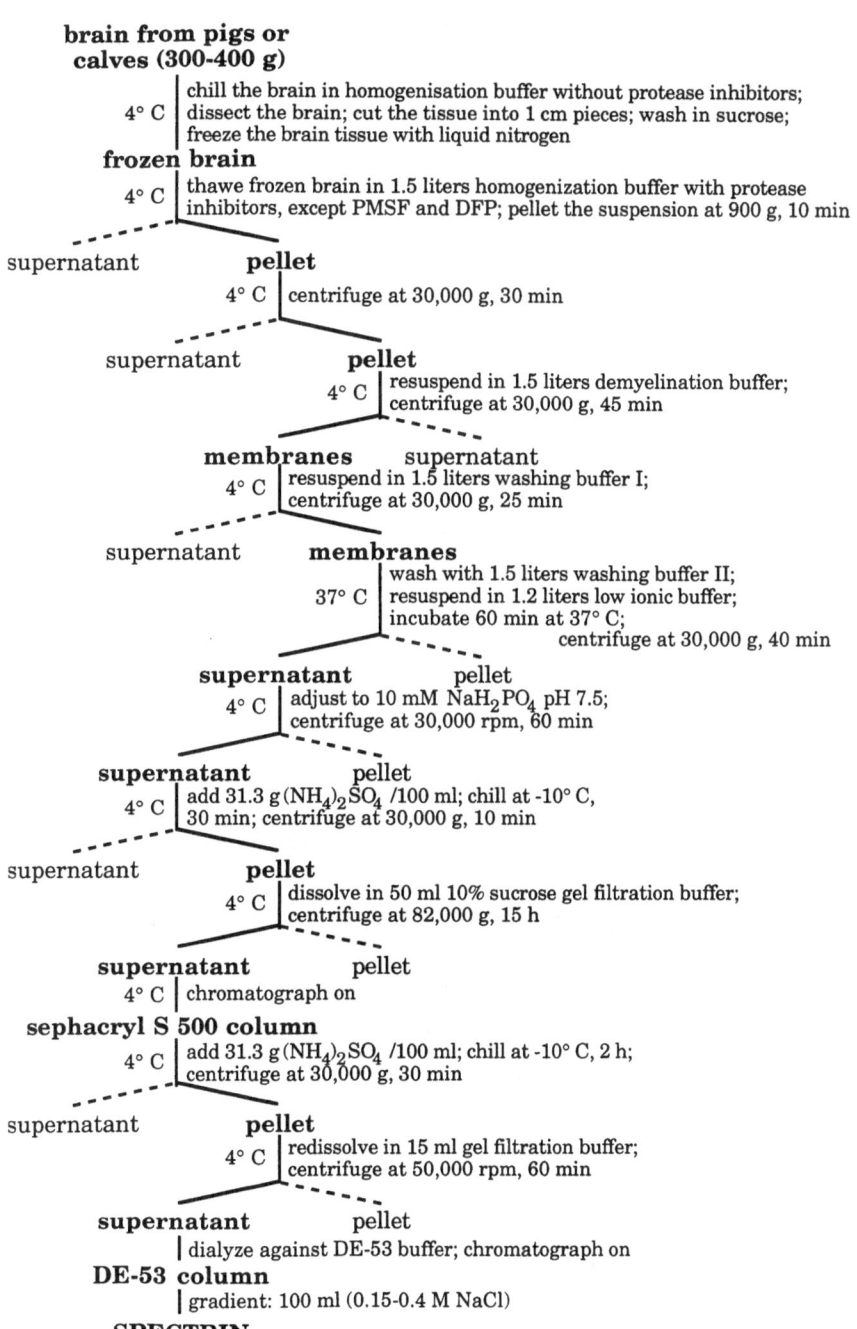

SYNAPSIN I

A 84 kDa phosphoprotein associated with synaptic vesicles. It binds to phospholipids and bundles F-actin in vitro in a phosphorylation dependent manner.

Source: brain from pigs or calves

Equipment:
- polytron-homogenizer (large head)
- centrifuge
- syringe without needle 50 cm^3
- CM 32-cellulose column (20 ml)
- hydroxylapatide column (2 ml)
- HPLC Mono S column
- electrophoretic gels
- chemical hood

Chemicals: sucrose, EGTA, EDTA, NaH$_2$PO$_4$, PIPES, DTT, NaCl, KCl, Tween 20, Triton X-100, KJ, NaBr, NaN$_3$, (NH$_4$)$_2$SO$_4$ (up), DMSO,

protease inhibitors: leupeptin, pepstatin, PMSF, DFP

Have ready: **homogenization buffer:** 0.32 M sucrose, 2 mM EGTA

demyelination buffer: 1 M sucrose, 2 mM EGTA, 100 mg/ml PMSF, pH 7.5

wash buffer I: 10 mM NaH$_2$PO$_4$, 100 mg/ml PMSF, pH 7.5

synapsin preextraction buffer: 10 mM NaH$_2$PO$_4$, 2 mM EGTA, 1% Triton X-100 (v/v), 2 mM DTT

synapsin extraction buffer: final concentration of 0.15 M NaCl, 10 mM NaH$_2$PO$_4$, 2 mM EGTA, 0.1% Triton X-100 (v/v), 2 mM DTT

CM-53 column-buffer: 10 mM PIPES, 1 mM EDTA, 0.5 mM DTT, 0.05% (v/v) Tween 20, 1 mM NaN$_3$, pH 6.8

hydroxylapatide column buffer: CM-53 column buffer plus 0.1 M NaCl, plus 2 mM MgCl$_2$, pH 6.8

storage buffer: 20% (w/v) sucrose, 0.1 M PIPES, 1 mM EGTA, 1 mM DTT, 1 mM NaN$_3$, pH 6.6

protease inhibitors: 5 mg/ml leupeptin, 5 mg/ml pepstatin, dissolved first in 0.5 ml DMSO, 200 mg/ml PMSF dissolved first at 200 mg/ml in DMSO, DFP 0.02% (v/v), pH 7.5

Reference: Bähler, M. and Greengard, P. (1987) Nature 326, 704-707

Gen-Bank: M27812 (rat), M27810 (bovine), J05431 (human)

brain from pigs or calves (300-400 g)

4° C | chill the brain in homogenization buffer without protease inhibitors; dissect the brain; cut the tissue into 1 cm pieces; wash in sucrose; freeze the brain tissue with liquid nitrogen

frozen brain

4° C | thawe frozen brain in 1.5 liters homogenization buffer with protease inhibitors, except PMSF and DFP; pellet the suspension at 900 g, 10 min

supernatant   **pellet**

4° C | centrifuge at 30,000 g, 30 min

supernatant   **pellet**

4° C | resuspend in 1.5 liters demyelination buffer; centrifuge at 30,000 g, 45 min

supernatant   **membranes**

4° C | resuspend in 1.5 liters wash buffer I; centrifuge at 30,000 g, 25 min

demyelinated membranes   supernatant

0-4° C | resuspend to 1.5 liters synapsin preextraction buffer; incubate 15 min on ice; centrifuge at 30,000 g

pellet   supernatant

4° C | redissolve the pellet in 300 ml synapsin extraction buffer; dialyze overnight against 4 liters of CM 32 column buffer; adsorb batchwise with 20 ml CM-cellulose equilibrated with the same buffer; pack the column; wash with 0.05 M NaCl in column buffer; elute with 0.1 M NaCl, 2 mM $MgCl_2$ directly on

hydroxyapatide column

synapsin is in the effluent; chromatograph on

HPLC Mono S column

4° C | elute with a linear gradient 0-0.3 M NaCl; dialyze against synapsin storage buffer

SYNAPSIN I

TALIN

Talin, a 270 kDa lipid and actin binding phosphoprotein that nucleates actin polymerization and links microfilaments to plasma membranes. 2541 amino acids, dumbbell-shaped homodimers of 51 nm contour length.

Source: chicken gizzard smooth muscle

Equipment :
- sharp knife
- waring blender
- glass-teflon-homogenizer with pestle
- DEAE-cellulose column (2.5 x 15 cm)
- sepharose 6B column (2.5 x 90 cm)
- hydroxylapatite column (1.5 x 15 cm)
- Amicon cell
- electrophoretic gels
- centrifuge

Chemicals: NaCl, Trizma Base, EGTA, EDTA, KH_2PO_4, $MgCl_2$, $(NH_4)_2SO_4$ (up), NaN_3, 2-mercaptoethanol
protease inhibitors: PMSF
column materials: sepharose 6B, DEAE-cellulose, hydroxylapatite

Have ready: **buffer A:** 2 mM Tris-HCl, 1 mM EGTA, 1 mM EDTA, 0.5 mM PMSF, pH 9.0 (1 liter)

buffer B (column buffer): 20 mM Tris-HCl, 0.1 mM EGTA, 0.1 mM EDTA, 20 mM NaCl, 0.1% 2-mercaptoethanol, pH 7.6 (10 liters)

phosphate buffer (hydroxylapatide column): 0.1 mM EGTA, 0.1% 2-mercaptoethanol, 0.1 M KH_2PO_4, pH 7.0 (3 liters)

Reference: O'Halloran, T., Molony, L. & Burridge, K. (1986) Methods Enzymol. 134, 69-77
Goldmann, W.H., Bremer, A., Häner, M., Aebi, U., and Isenberg, G. (1994) J. Struct. Biol. 112, 3-10
Gen-Bank: X56123 (mouse fibroblast)

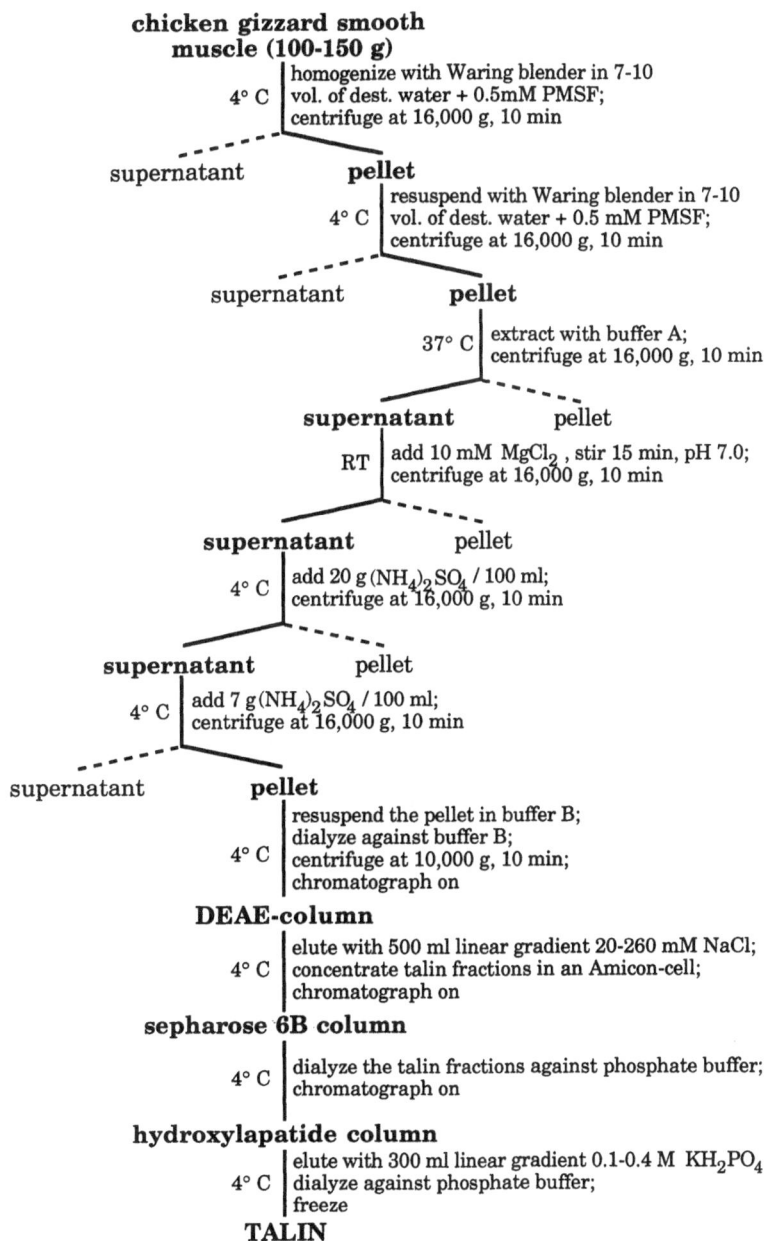

chicken gizzard smooth muscle (100-150 g)

4° C | homogenize with Waring blender in 7-10 vol. of dest. water + 0.5mM PMSF; centrifuge at 16,000 g, 10 min

supernatant — **pellet**

4° C | resuspend with Waring blender in 7-10 vol. of dest. water + 0.5 mM PMSF; centrifuge at 16,000 g, 10 min

supernatant — **pellet**

37° C | extract with buffer A; centrifuge at 16,000 g, 10 min

supernatant — pellet

RT | add 10 mM $MgCl_2$, stir 15 min, pH 7.0; centrifuge at 16,000 g, 10 min

supernatant — pellet

4° C | add 20 g $(NH_4)_2SO_4$ / 100 ml; centrifuge at 16,000 g, 10 min

supernatant — pellet

4° C | add 7 g $(NH_4)_2SO_4$ / 100 ml; centrifuge at 16,000 g, 10 min

supernatant — **pellet**

4° C | resuspend the pellet in buffer B; dialyze against buffer B; centrifuge at 10,000 g, 10 min; chromatograph on

DEAE-column

4° C | elute with 500 ml linear gradient 20-260 mM NaCl; concentrate talin fractions in an Amicon-cell; chromatograph on

sepharose 6B column

4° C | dialyze the talin fractions against phosphate buffer; chromatograph on

hydroxylapatide column

4° C | elute with 300 ml linear gradient 0.1-0.4 M KH_2PO_4 dialyze against phosphate buffer; freeze

TALIN

TENSIN

A 170 kDa polypeptide binding to the barbed ends of actin filaments and to vinculin. Localized at actin/membrane linkage structures. Presence of phosphotyrosine and SH2-domains.

Source:	chicken gizzard
Equipment:	• Waring blender
	• water bath
	• DEAE-cellulose column suspension (100 ml)
	• DEAE-cellulose (DE-53) column (2.5 x 50 cm)
	• hydroxylapatide column (1.3 x11 cm)
Chemicals:	Tris-HCl, EDTA, EGTA, Tris-acetate, NaCl, 2-mercaptoethanol, $(NH_4)_2SO_4$
	protease inhibitors: PMSF
Have ready: EDTA,	**buffer A:** 2 mM Tris-HCl, 1 mM EGTA, 1 mM 0.5 mM PMSF, pH 9.0
	buffer B: 20 mM Tris-acetate, 20 mM NaCl, 15 mM 2-mercaptoethanol, 0.1 mM EDTA, pH 7.6
	phosphate buffer, 50 mM, pH 7.6
Reference:	Wilkins, J.A., Risinger, M.A. and Lin, S. (1986) J. Cell Biol. 103, 1483-1494

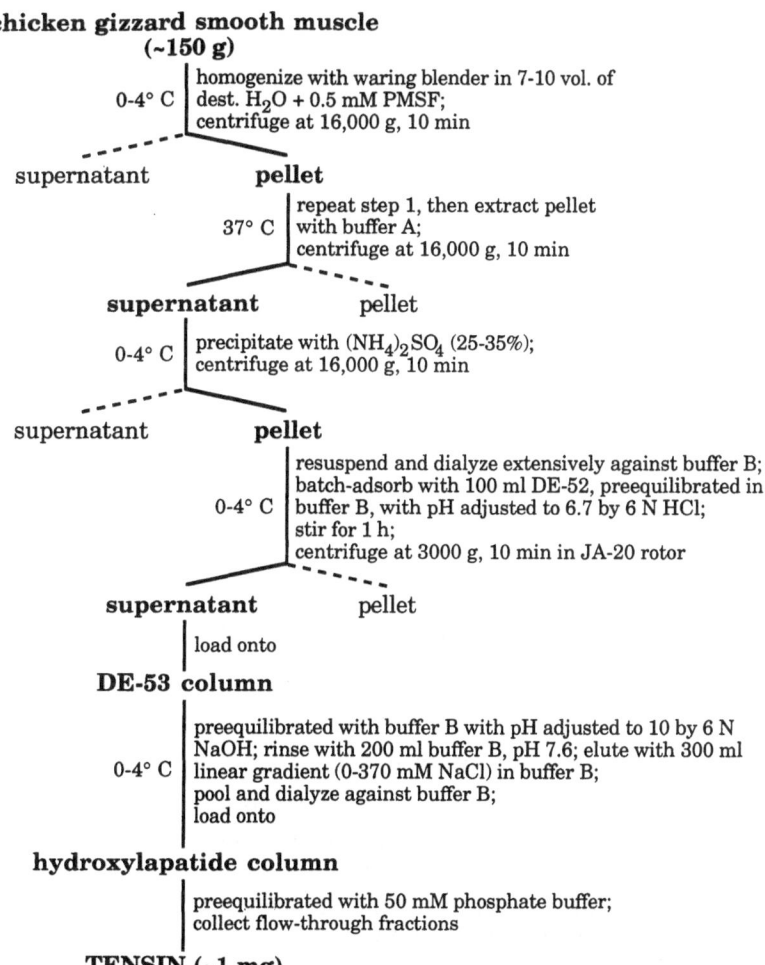

chicken gizzard smooth muscle
(~150 g)

0-4° C | homogenize with waring blender in 7-10 vol. of dest. H_2O + 0.5 mM PMSF; centrifuge at 16,000 g, 10 min

supernatant   **pellet**

37° C | repeat step 1, then extract pellet with buffer A; centrifuge at 16,000 g, 10 min

supernatant   pellet

0-4° C | precipitate with $(NH_4)_2SO_4$ (25-35%); centrifuge at 16,000 g, 10 min

supernatant   **pellet**

0-4° C | resuspend and dialyze extensively against buffer B; batch-adsorb with 100 ml DE-52, preequilibrated in buffer B, with pH adjusted to 6.7 by 6 N HCl; stir for 1 h; centrifuge at 3000 g, 10 min in JA-20 rotor

supernatant   pellet

load onto

DE-53 column

0-4° C | preequilibrated with buffer B with pH adjusted to 10 by 6 N NaOH; rinse with 200 ml buffer B, pH 7.6; elute with 300 ml linear gradient (0-370 mM NaCl) in buffer B; pool and dialyze against buffer B; load onto

hydroxylapatide column

preequilibrated with 50 mM phosphate buffer; collect flow-through fractions

TENSIN (~1 mg)

THYMOSIN β4

A 5 kDa peptide, widely distributed in vertebrate cells. The most
abundant actin sequestering protein in human platelets. A single
polypeptide of 43 amino acids which binds to G-actin.

Source: calf thymus

Equipment:
- Waring blender
- glass wool
- miracloth
- Buchner funnel
- ultra filtration (Amicon DC-2 hollow fiber system) H1 DP10 membrane cartridge
- sephadex G-25 column (5 x 80 cm)
- rotary evaporation
- DE-52 cellulose column (3.8 x 50 cm)
- sephadex G-10 column (5 x 80 cm)
- sephadex G-75 column (1.8 x 150 cm)

Chemicals: NaCl, octyl alcohol, acetone (99.5%), acetic acid, (10%), saturated $(NH_4)_2SO_4$, solid $(NH_4)_2SO_4$, Tris-HCl, 2-mercaptoethanol, guanidinium chloride, DTT, Na/PO_4, ammonium bicarbonate, NaOH

Have ready: **10 mM sodium phosphate buffer,** pH 7.0

0.1 M ammonium bicarbonate

buffer A: 10 mM Tris-HCl, 1 mM 2-mercaptoethanol, 0.1 mM DTT, pH 8.5

dissolving buffer: 6 M guanidinium chloride, 10 mM Tris-HCl, pH 7.5

Reference: Low, T. L. K. and Goldstein, A. L. (1985) Methods Enzymol. 116, 219-233 and 248-255

frozen calf thymus (~2 kg)

4° C | homogenize in 6 l 0.15 M NaCl with Waring blender; add 100 ml octylalcohol to minimize foaming; centrifuge at 14,000 g, 20 min

supernatant   pellet

filter through glasswool; heat up to 80°C (2 l batches) and filter through miracloth into ice-cooled flasks;
add to 5 vol. acetone (99.5%) at -10°C; collect precipitate on Buchner funnel; wash several times with acetone (-10°C) and dry under vacuum

acetone powder

4° C | dissolve in 10 vol (8/10 ml) sodium phosphate buffer; stir at RT, 30 min; centrifuge at 15,000 g, 30 min;

supernatant   pellet

4° C | adjust protein concentration to 25 mg/ml as determined by Lowry; add 33.3 ml saturated $(NH_4)_2SO_4$ (25%) (pH adjusted to 7.0 with NaOH) dropwise to 100 ml supernatant; stir for 30 min; centrifuge at 15,000 g, 15 min

supernatant   pellet

4° C | adjust pH to 4.0 with 10% acetic acid; add solid $(NH_4)_2SO_4$, 14.6 g/100 ml (50%) and stir 1 h or overnight; centrifuge at 15,000 g, 30 min

supernatant   pellet

4° C | add solid $(NH_4)_2SO_4$, 30.8 g/100 ml (95%) and stir for 1 h or longer; centrifuge at 15,000 g, 30 min

supernatant   **pellet**

dissolve in 0.1 M ammonium bicarbonate to 10 mg/ml protein and subject to ultrafiltration; concentrate by rotary evaporation and desalt on sephadex G-25 column equilibrated with sterile, distilled H_2O; pool protein peak; concentrate and lyophilize

F5A (thymosin fraction) ~0.7g/1 kg wet tissue

F5A thymosin fractions (~3 g)

dissolve in 200 ml buffer A and stir overnight; apply to DEAE cellulose column equilibrated with buffer A and elute with a linear gradient (2 l each) of 0-0.4 M NaCl;
elute a second time with a linear gradient (2 l each) of 0.4-0.8 M NaCl, monitore at 270 nm; pool the second peak (0.045 M NaCl)

thymosin fractions

desalt on sephadex G-10 column; lyophilize (total protein 150 mg); 70 mg of protein is dissolved in 2 ml dissolving buffer; add 14 µl 2-mercaptoethanol; stir overnight at RT; apply to sephadex G-75 column equilibrated with dissolving buffer; monitore effluents at 235 nm; pool the fourth protein peak; desalt on sephadex G-10 column;

THYMOSIN β4 (20-25 mg)

131

TITIN

A giant polypeptide of 2000-3500 kDa. Building blocks of highly flexible strands (1 μm long with 3-4 nm in diameter). Probably providing an elastic restoring force after muscle contraction.

Source:	skeletal muscle
Equipment:	• Waring blender • centrifuge, GSA rotor • high porosity SDS-gel system (see nebulin purification) • water bath • Bio Gel A-150 M column (2.5 x 90 cm)
Chemicals:	KCl, MgCl$_2$, EGTA, Tris-maleate, PMSF, DTT, Triton X-100, SDS, sodium acetate, Tris-HCl, EDTA, Na$_4$P$_2$O$_7$, NaCl
Have ready:	**PRB-buffer (pyrophosphate relaxing buffer):** 0.1 M KCl, 2 mM MgCl$_2$, 2 mM EGTA, 10 mM Tris-maleate, 0.5 mM DTT, 0.1 mM PMSF, 2 mM Na$_4$P$_2$O$_7$, pH 6.8
	low salt buffer: PRB-buffer without pyrophosphate
	Triton-buffer: low salt buffer + 0.5% Triton X-100
SDS,	**washing buffer:** 5 mM Tris-HCl, 4 M NaCl, 3% pH 8.0
	gelfiltration column buffer: 40 mM Tris-HCl, 20 mM sodium acetate, 2 mM EDTA, 0.1 mM DTT, 0.1% SDS, pH 7.4
Reference:	Wang, K. (1982) Methods Enzymol. 85, 264-274 Gen-Bank: X17329 and X17330 (rabbit), X15423 (nematode twitchin), M73433 (Lethocerous projectin), M73435 (drosophile projectin)

rabbit back muscles

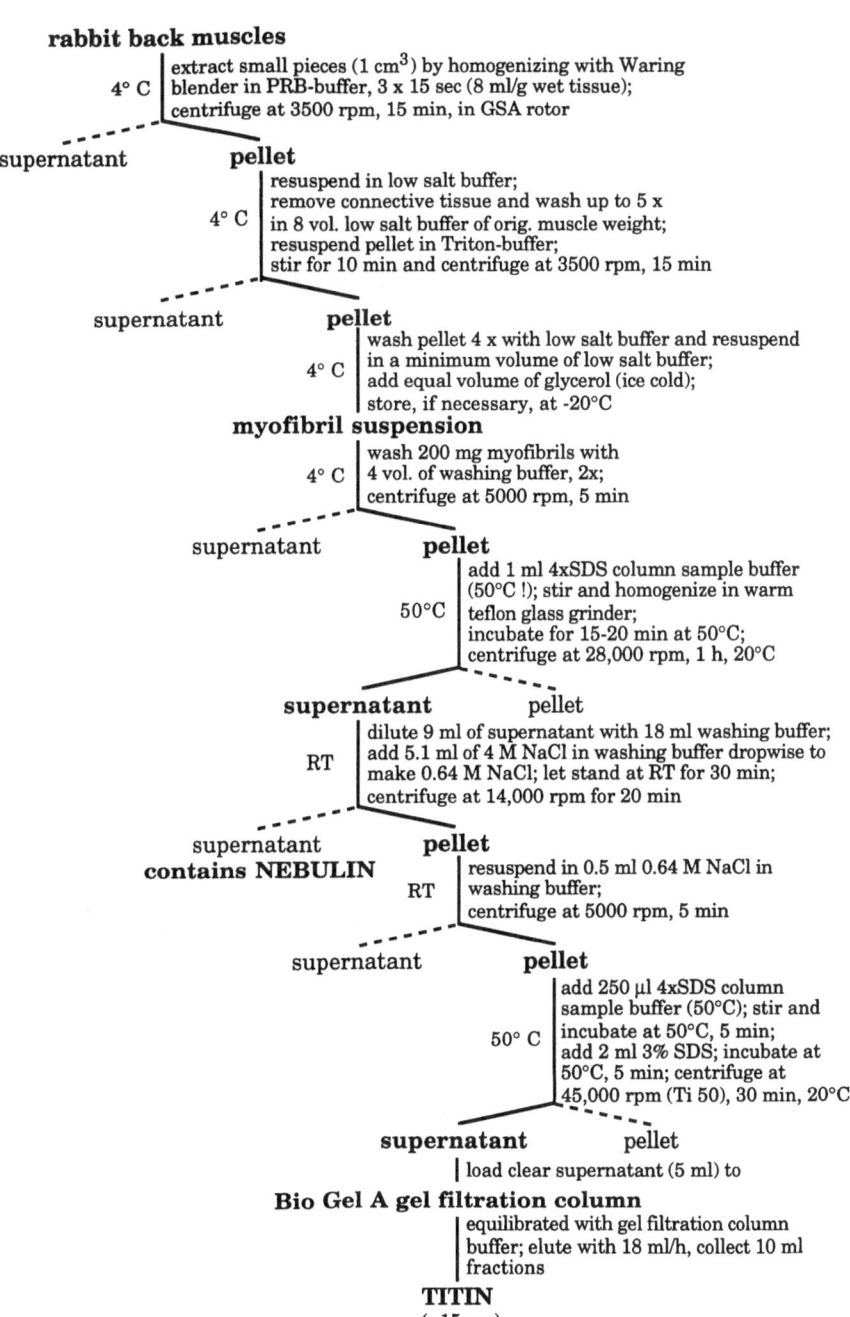

4° C — extract small pieces (1 cm³) by homogenizing with Waring blender in PRB-buffer, 3 x 15 sec (8 ml/g wet tissue); centrifuge at 3500 rpm, 15 min, in GSA rotor

supernatant   **pellet**

4° C — resuspend in low salt buffer; remove connective tissue and wash up to 5 x in 8 vol. low salt buffer of orig. muscle weight; resuspend pellet in Triton-buffer; stir for 10 min and centrifuge at 3500 rpm, 15 min

supernatant   **pellet**

4° C — wash pellet 4 x with low salt buffer and resuspend in a minimum volume of low salt buffer; add equal volume of glycerol (ice cold); store, if necessary, at -20°C

myofibril suspension

4° C — wash 200 mg myofibrils with 4 vol. of washing buffer, 2x; centrifuge at 5000 rpm, 5 min

supernatant   **pellet**

50°C — add 1 ml 4xSDS column sample buffer (50°C !); stir and homogenize in warm teflon glass grinder; incubate for 15-20 min at 50°C; centrifuge at 28,000 rpm, 1 h, 20°C

supernatant   pellet

RT — dilute 9 ml of supernatant with 18 ml washing buffer; add 5.1 ml of 4 M NaCl in washing buffer dropwise to make 0.64 M NaCl; let stand at RT for 30 min; centrifuge at 14,000 rpm for 20 min

supernatant   **pellet**
contains NEBULIN

RT — resuspend in 0.5 ml 0.64 M NaCl in washing buffer; centrifuge at 5000 rpm, 5 min

supernatant   **pellet**

50° C — add 250 µl 4xSDS column sample buffer (50°C); stir and incubate at 50°C, 5 min; add 2 ml 3% SDS; incubate at 50°C, 5 min; centrifuge at 45,000 rpm (Ti 50), 30 min, 20°C

supernatant   pellet

load clear supernatant (5 ml) to

Bio Gel A gel filtration column

equilibrated with gel filtration column buffer; elute with 18 ml/h, collect 10 ml fractions

TITIN
(~15 mg)

TRANSGELIN

A wide spread, monomeric, 21 kDa (122 amino acids) transformation sensitive actin gelling protein with high sequence homology of 22 kDa avian SM22a protein. Not present in skeletal muscle, erythrocytes and neurons.

Source:	sheep aorta
Equipment:	• Waring blender • glass wool • SDS-PAGE • chromatofocusing column pH 9-7 (1.6 x 70 cm) • hydroxylapatide column (1 x 1 cm)
Chemicals:	KCl, MgCl$_2$, imidazole, NaN$_3$, EGTA, EDTA, DTT, Na-acetate, CHAPS **protease inhibitors:** leupeptin, chymostatin, pepstatin, PMSF
Have ready:	**buffer A:** 60 mM KCl, 4 mM MgCl$_2$, 10 mM imidazole, 1 mM NaN$_3$, 0.5 mM EGTA, 5 mM DTT, 0.1 mM EDTA, 0.2 mM PMSF + 2 µg/ml protease inhibitors, pH 7.1
	buffer B: 50 mM Na-acetate, 4 mM MgCl$_2$, 0.5 M KCl, pH 4.8
	ethanolamine-HCl, pH 9.6
	phosphate buffer, pH 6.8
Reference:	Shapland, C. et al. (1993) J. Cell Biol. 121, 1065-1073

sheep aorta (10 g)

4° C │ stored at -196° C; fracture into small pieces,
     │ homogenize with blender in 5 vol. buffer A +
     │ 0.5% CHAPS (3 x 10s top speed); leave on
     │ ice for 1 h; centrifuge at 23,000 g, 5 min

supernatant   pellet

│ filter through glass wool;
│ dialyze against 2 l buffer B, 16 h;
│ clarify at 48,000 g, 2 h;
│ dialyze against 25 mM ethanolamine-HCl, pH 9.6;
│ load onto

chromatofocusing column

│ collect transgelin fractions identified by SDS-PAGE;
│ dialyze overnight against 20 mM phosphate buffer, pH 6.8;
│ apply to

hydroxylapatide column

│ preequilibrate with 20 mM phosphate buffer;
│ elute with 100 mM phosphate buffer

TRANSGELIN
(~1 mg)

TROPOMODULIN

A globular 43 kDa tropomyosin binding protein which blocks self-association and inhibits F-actin binding of tropomyosin. 359 amino acids, predominantly α-helical, blocks polymerization at the pointed (-) end of actin filaments.

Source: erythrocytes

Equipment:
- Leuko-Pak filter (Feuwal-Lab. Inc., Deerfield Il.)
- Millipore filter system
- centrifuge
- DEAE-cellulose column (DE-53) (2.5 x 18 cm)
- hydroxylapatide column (0.7 x 2 cm)
- vacuum dialysis equipment
- Ultrogel AcA 44 gel filtration column (0.9 x 54 cm)

Chemicals: NaCl, Na/PO$_4$, dextran T-500, MgCl$_2$, EGTA, EDTA, DTT, Tris-HCl, KCl, HEPES, NaN$_3$, K/PO$_4$, (NH$_4$)$_2$SO$_4$

protease inhibitor stock solution in DMSO: TPCK (N-tosyl-L-phenylalanine chloromethylketone) and PMSF (20 mg/ml), 1,10-phenanthroline, leupeptin, pepstatin A, chymostatin, antipain, benzamidine (10 mg/ml), DFP, 200 mg/ml PMSF in DMSO

Have ready: **washing solution I:** 150 mM NaCl, 5 mM sodium phosphate, pH 7.5

lysis buffer: 7.5 mM Na/PO$_4$, 2 mM MgCl$_2$, 1 mM EGTA, 2 mM DTT, 20 µg/ml PMSF, pH 7.5

washing solution II: 7.5 mM Na/PO$_4$, 1 mM EDTA, 2 mM DTT + 20 µg/ml PMSF, pH 7.5

2 M Tris-HCl buffer, pH 8.5

buffer II: 20 mM Tris-HCl, 1 mM EDTA, 2 mM DTT, pH 8.5

hydroxlyapatide column buffer: 10 mM K/PO$_4$, 1 mM DTT, pH 7.0

vacuum dialysis chamber buffer: 100 mM KCl, 20 mM HEPES, 0.2 mM EDTA, 2 mM DTT, 0.02% NaN$_3$, pH 7.3

Reference: Fowler, V.M. (1987) J. Biol. Chem. 262, 12792-12800
Gen-Bank: M77016 (human)

human blood (2 units)

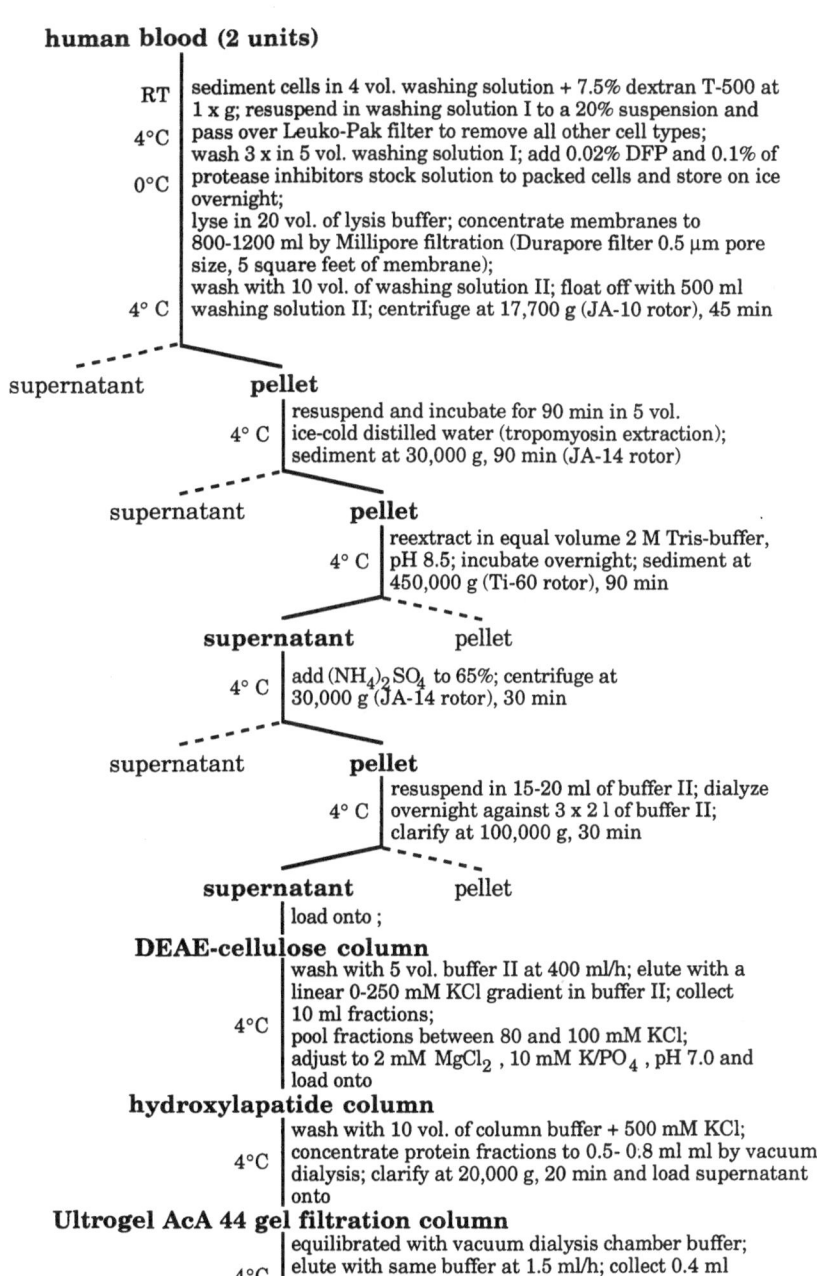

RT | sediment cells in 4 vol. washing solution + 7.5% dextran T-500 at 1 x g; resuspend in washing solution I to a 20% suspension and
4°C | pass over Leuko-Pak filter to remove all other cell types; wash 3 x in 5 vol. washing solution I; add 0.02% DFP and 0.1% of
0°C | protease inhibitors stock solution to packed cells and store on ice overnight;
lyse in 20 vol. of lysis buffer; concentrate membranes to 800-1200 ml by Millipore filtration (Durapore filter 0.5 μm pore size, 5 square feet of membrane);
wash with 10 vol. of washing solution II; float off with 500 ml
4° C | washing solution II; centrifuge at 17,700 g (JA-10 rotor), 45 min

supernatant

pellet

4° C | resuspend and incubate for 90 min in 5 vol. ice-cold distilled water (tropomyosin extraction); sediment at 30,000 g, 90 min (JA-14 rotor)

supernatant

pellet

4° C | reextract in equal volume 2 M Tris-buffer, pH 8.5; incubate overnight; sediment at 450,000 g (Ti-60 rotor), 90 min

supernatant   pellet

4° C | add $(NH_4)_2SO_4$ to 65%; centrifuge at 30,000 g (JA-14 rotor), 30 min

supernatant

pellet

4° C | resuspend in 15-20 ml of buffer II; dialyze overnight against 3 x 2 l of buffer II; clarify at 100,000 g, 30 min

supernatant   pellet

load onto ;

DEAE-cellulose column

4°C | wash with 5 vol. buffer II at 400 ml/h; elute with a linear 0-250 mM KCl gradient in buffer II; collect 10 ml fractions;
pool fractions between 80 and 100 mM KCl; adjust to 2 mM $MgCl_2$, 10 mM K/PO_4, pH 7.0 and load onto

hydroxylapatide column

4°C | wash with 10 vol. of column buffer + 500 mM KCl; concentrate protein fractions to 0.5- 0.8 ml ml by vacuum dialysis; clarify at 20,000 g, 20 min and load supernatant onto

Ultrogel AcA 44 gel filtration column

4°C | equilibrated with vacuum dialysis chamber buffer; elute with same buffer at 1.5 ml/h; collect 0.4 ml fractions; pool, concentrate to 0.2 mg/ml by vacuum dialysis and store at 0°C on ice

TROPOMODULIN

TROPOMYOSIN(s)

An ubiquitous actin binding protein arranged in α-helical coiled-coil of 410 Å length.

Source: rabbit skeletal muscle

Equipment:
- hydroxylapatide column (2.5 x 25 cm)
- electrophoretic gels
- centrifuge

Chemicals: KCl, Trizma Base, KH_2PO_4, DL-DTT, 2-mercaptoethanol, imidazole, $(NH_4)_2SO_4$
column materials: hydroxylapatide

Have ready: **extraction buffer:** 10 mM Tris-HCl, 0.5 mM 2-mercaptoethanol, pH 8.0 (200 ml)

imidazole buffer: 2 mM imidazole, 5 mM 2-mercaptoethanol, pH as made (2 liters)

phosphate buffer: 1 M KCl. 1mM KH_2PO_4, 2 mM DTT, pH 7.0 (5 liters)

Reference: Smillie, L.B. (1982) Methods Enzymol. 85, 234-241

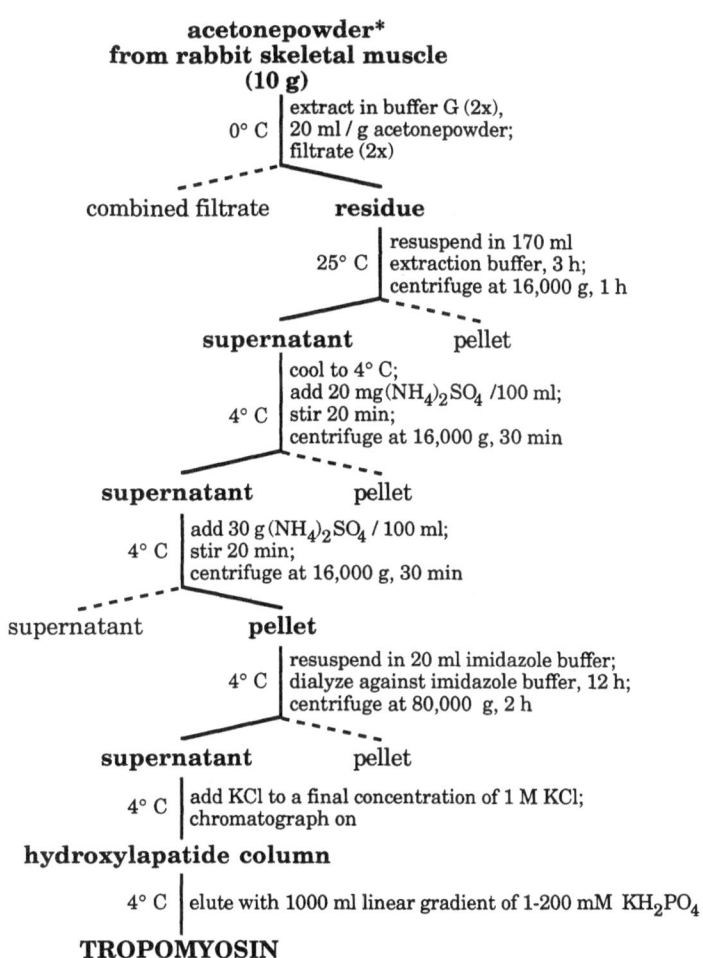

acetonepowder*
from rabbit skeletal muscle
(10 g)

0° C | extract in buffer G (2x),
20 ml / g acetonepowder;
filtrate (2x)

combined filtrate   **residue**

25° C | resuspend in 170 ml
extraction buffer, 3 h;
centrifuge at 16,000 g, 1 h

supernatant   pellet

4° C | cool to 4° C;
add 20 mg $(NH_4)_2SO_4$ /100 ml;
stir 20 min;
centrifuge at 16,000 g, 30 min

supernatant   pellet

4° C | add 30 g $(NH_4)_2SO_4$ / 100 ml;
stir 20 min;
centrifuge at 16,000 g, 30 min

supernatant   **pellet**

4° C | resuspend in 20 ml imidazole buffer;
dialyze against imidazole buffer, 12 h;
centrifuge at 80,000 g, 2 h

supernatant   pellet

4° C | add KCl to a final concentration of 1 M KCl;
chromatograph on

hydroxylapatide column

4° C | elute with 1000 ml linear gradient of 1-200 mM KH_2PO_4

TROPOMYOSIN

---

* *See preparation of smooth muscle actin*

TROPONINS

TnC (17.8 kDa), TnI (20.8 kDa) and TnT (30.5 kDa) comprise the Ca^{++}-dependent regulatory protein complex located on the thin actin filaments of muscle. Involved in key regulatory mechanism for muscle contraction.

Source: skeletal muscle ether powder

Equipment:
- centrifuge
- Cibacron Blue-sephacryl column (5 x 50 cm)
- DEAE-sephadex A-50 column (2.5 x 25 cm)
- CM-sephadex C-50-120 column (2.5 x 25 cm)
- SDS-PAGE

Chemicals: Tris-HCl, KCl, $CaCl_2$, DTT, 1 N HCl, 1 N KOH, NaN_3, imidazole, urea, EDTA, citrate

Have ready: **extraction solution:** 1 M KCl, 25 mM Tris-HCl, 0.1 mM $CaCl_2$, 0.1 mM DTT, pH 8.0

dialyzing buffer 1: 10 mM imidazole, 50 mM KCl, 0.1 mM $CaCl_2$, 0.1 mM DTT, 0.02% NaN_3, pH 7.0

dialyzing buffer 2: 2 mM Tris-HCl, 0.1 mM $CaCl_2$, pH 8.0

dialyzing buffer 3: 6 M urea, 50 mM Tris-HCl, 1 mM EDTA, 0.1 mM DTT, pH 8.0

dialyzing buffer 4: 6 M urea, 50 mM citrate, 1 mM EDTA, 0.1 mM DTT, pH 6.0

Reference: Potter, J.D. (1982) Methods Enzymol. 85, 241-263
Gen-Bank: TnC Y00760, Y0346; TnI X54163

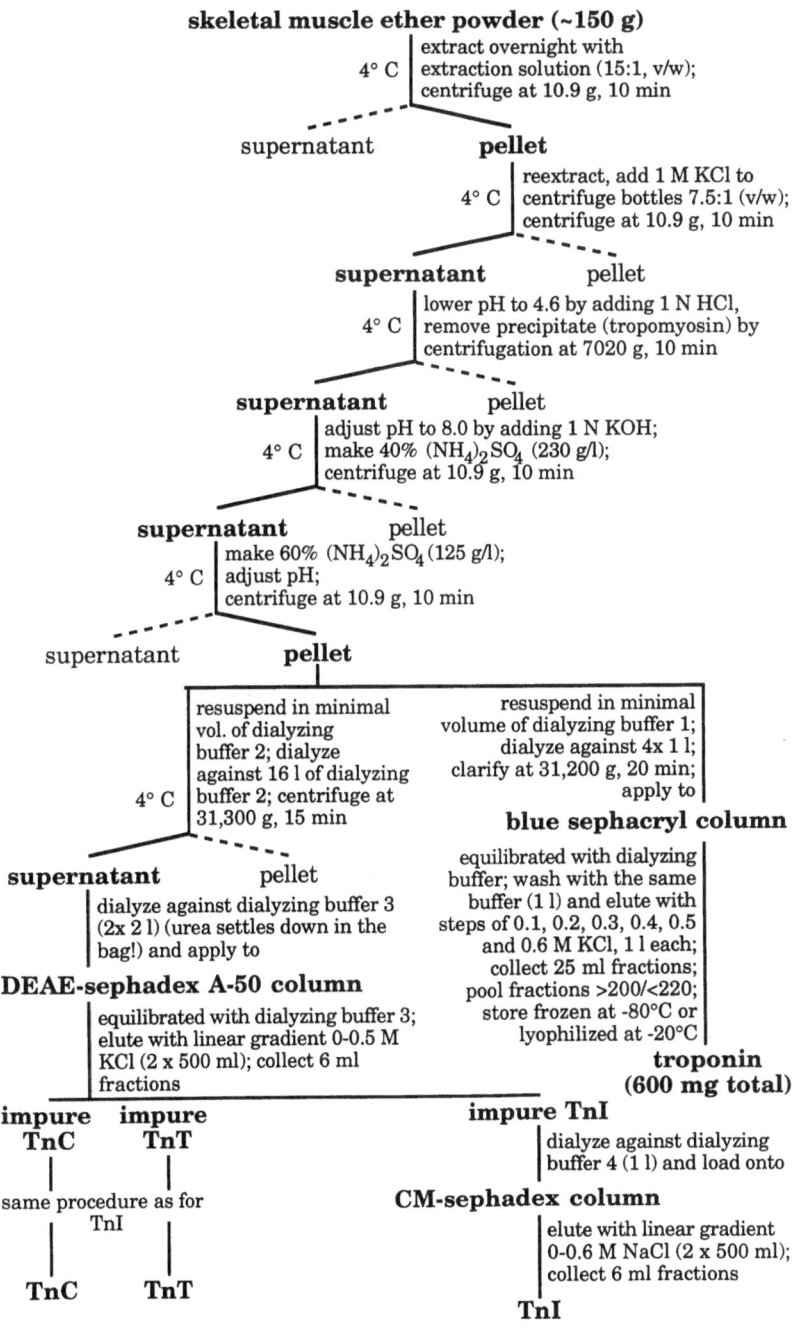

skeletal muscle ether powder (~150 g)

4° C | extract overnight with extraction solution (15:1, v/w); centrifuge at 10.9 g, 10 min

supernatant   **pellet**

4° C | reextract, add 1 M KCl to centrifuge bottles 7.5:1 (v/w); centrifuge at 10.9 g, 10 min

supernatant   pellet

4° C | lower pH to 4.6 by adding 1 N HCl, remove precipitate (tropomyosin) by centrifugation at 7020 g, 10 min

supernatant   pellet

4° C | adjust pH to 8.0 by adding 1 N KOH; make 40% $(NH_4)_2SO_4$ (230 g/l); centrifuge at 10.9 g, 10 min

supernatant   pellet

4° C | make 60% $(NH_4)_2SO_4$ (125 g/l); adjust pH; centrifuge at 10.9 g, 10 min

supernatant   **pellet**

4° C | resuspend in minimal vol. of dialyzing buffer 2; dialyze against 16 l of dialyzing buffer 2; centrifuge at 31,300 g, 15 min

resuspend in minimal volume of dialyzing buffer 1; dialyze against 4x 1 l; clarify at 31,200 g, 20 min; apply to **blue sephacryl column**

supernatant   pellet

dialyze against dialyzing buffer 3 (2x 2 l) (urea settles down in the bag!) and apply to

DEAE-sephadex A-50 column

equilibrated with dialyzing buffer 3; elute with linear gradient 0-0.5 M KCl (2 x 500 ml); collect 6 ml fractions

equilibrated with dialyzing buffer; wash with the same buffer (1 l) and elute with steps of 0.1, 0.2, 0.3, 0.4, 0.5 and 0.6 M KCl, 1 l each; collect 25 ml fractions; pool fractions >200/<220; store frozen at -80°C or lyophilized at -20°C

troponin (600 mg total)

impure TnC   **impure TnT**   **impure TnI**

dialyze against dialyzing buffer 4 (1 l) and load onto

CM-sephadex column

same procedure as for TnI

elute with linear gradient 0-0.6 M NaCl (2 x 500 ml); collect 6 ml fractions

TnC   **TnT**   **TnI**

VILLIN

A monomeric actin binding protein of 95 kDa with Ca^{++}-dependent, PIP-2 inhibited effects on filament polymerization.

Source:	small intestines brush borders from chickens
Equipment:	• DNase I immobilized on sepharose 4B column • DEAE-cellulose column (3 ml) • centrifuge • electrophoretic gels
Chemicals:	NaCl, KH$_2$PO$_4$, NaH$_2$PO$_4$, KCl, EGTA, imidazole, MgCl$_2$, CaCl$_2$, Trizma-Base, DTT, glycerol, NaN$_3$, (NH$_4$)$_2$SO$_4$ (up) **protease inhibitors:** PMSF, benzamidine **column materials:** DEAE-cellulose
Have ready:	**buffer A:** 5 mM CaCl$_2$, 1 mM NaN$_3$, 10 mM Tris-HCl, pH 7.5 **buffer C:** 0.6 M KCl, 1 mM CaCl$_2$, 1 mM DTT, 10 mM Tris-HCl, pH 7.8 **buffer H:** 75 mM KCl, 1 mM EGTA, 0.1 mM MgCl$_2$, 0.25 mM PMSF, 0.5 mM benzamidine, 10 mM imidazole-HCl, pH 7.3 **buffer I:** 0.2 M KCl, 1 mM CaCl$_2$, 1 mM DTT, 10 mM imidazole-HCl, pH 7.3 **buffer J:** 0.15 M NaCl, 5 mM EGTA, 1 mM DTT, 10 mM Tris-HCl, pH 7.8 **buffer K:** 30 mM NaCl, 1 mM DTT, 20 mM Tris-HCl, pH 7.8 **storage buffer:** 30 mM KCl, 0.1 mM MgCl$_2$, 1 mM EGTA, 0.1 mM DTT, 10 mM imidazole-HCl, 50% glycerol, pH 7.3
Reference:	Bretscher, A. (1986) Methods Enzymol. 134, 24-32 Gen-Bank: X12901 (human), J03781 (chicken)

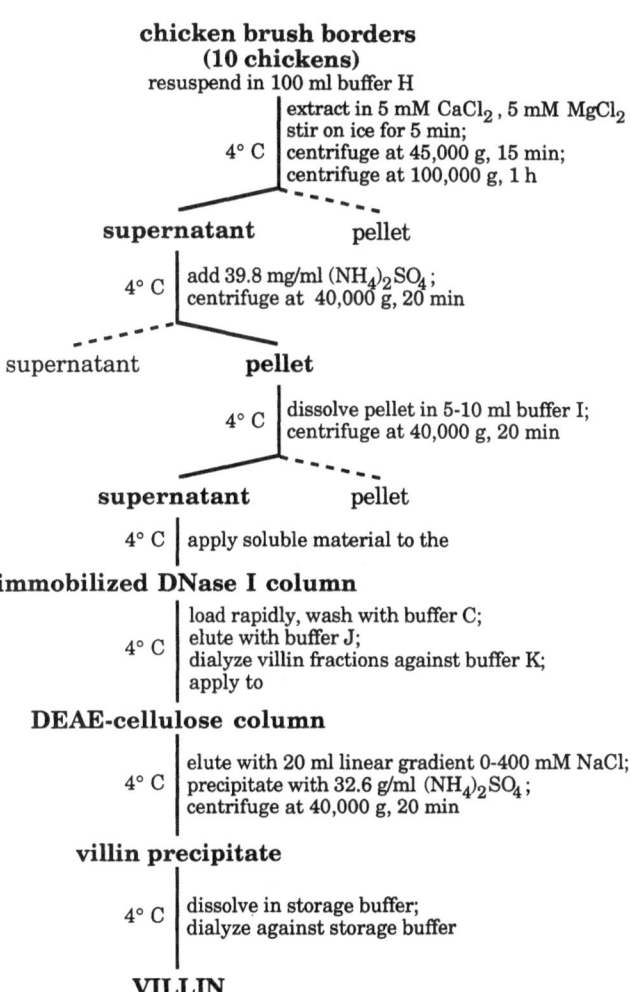

**chicken brush borders
(10 chickens)**
resuspend in 100 ml buffer H

4° C | extract in 5 mM $CaCl_2$, 5 mM $MgCl_2$;
stir on ice for 5 min;
centrifuge at 45,000 g, 15 min;
centrifuge at 100,000 g, 1 h

supernatant pellet

4° C | add 39.8 mg/ml $(NH_4)_2SO_4$;
centrifuge at 40,000 g, 20 min

supernatant **pellet**

4° C | dissolve pellet in 5-10 ml buffer I;
centrifuge at 40,000 g, 20 min

supernatant pellet

4° C | apply soluble material to the

immobilized DNase I column

4° C | load rapidly, wash with buffer C;
elute with buffer J;
dialyze villin fractions against buffer K;
apply to

DEAE-cellulose column

4° C | elute with 20 ml linear gradient 0-400 mM NaCl;
precipitate with 32.6 g/ml $(NH_4)_2SO_4$;
centrifuge at 40,000 g, 20 min

villin precipitate

4° C | dissolve in storage buffer;
dialyze against storage buffer

VILLIN
(~ 10 mg)

143

VINCULIN

A 117 kDa actin and lipid binding protein. A prominent linker protein of cellular junctions and focal adhesions.

Source: chicken gizzard smooth muscle

Equipment :
- sharp knife
- Waring blender
- glass-teflon-homogenizer with pestle
- DEAE-cellulose column (2.5 x 15 cm)
- sepharose 6B column (2.5 x 90 cm)
- electrophoretic gels
- centrifuge

Chemicals: NaCl, Trizma Base, EGTA, EDTA, Na-acetate, $MgCl_2$, $(NH_4)_2SO_4$ (up), NaN_3, 2-mercaptoethanol
protease inhibitors: PMSF
column materials: sepharose 6B, DEAE cellulose, CM-52-cellulose

Have ready: **buffer A:** 2 mM Tris-HCl, 1 mM EGTA, 1 mM EDTA,   0.5 mM PMSF, pH 9.0 (1 liter)

buffer B (column buffer): 20 mM Tris-HCl, 0.1 mM EGTA, 0.1 mM EDTA, 20 mM NaCl, 0.1% 2-mercaptoethanol, pH 7.6 (10 liters)

buffer C: 20 mM NaCl, 0.1% 2-mercaptoethanol, 0.1 mM EDTA, 20 mM Na-acetate, pH 5.0

Reference: Geiger, B. (1979) Cell 18, 193-205
O´Halloran, T., Molony, L. and Burridge, K. (1986) Methods Enzymol. 134, 69-77

**chicken gizzard smooth
muscle (100-150 g)**

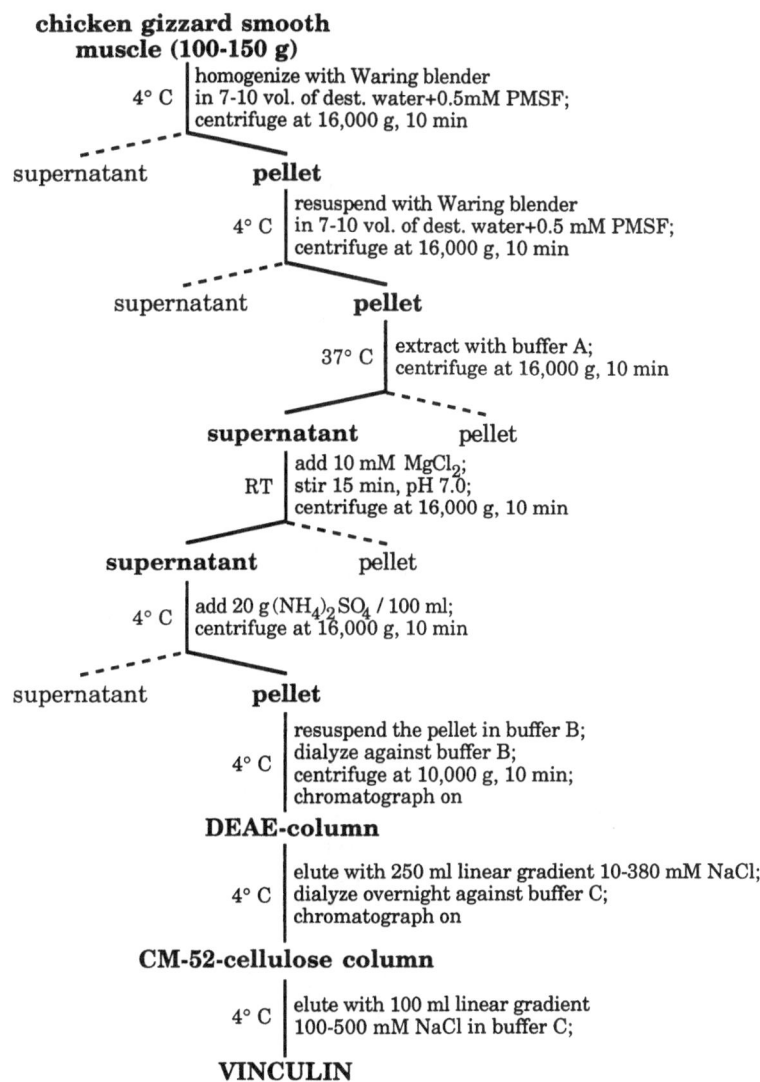

4° C | homogenize with Waring blender
in 7-10 vol. of dest. water+0.5mM PMSF;
centrifuge at 16,000 g, 10 min

supernatant   **pellet**

4° C | resuspend with Waring blender
in 7-10 vol. of dest. water+0.5 mM PMSF;
centrifuge at 16,000 g, 10 min

supernatant   **pellet**

37° C | extract with buffer A;
centrifuge at 16,000 g, 10 min

supernatant   pellet

RT | add 10 mM MgCl$_2$;
stir 15 min, pH 7.0;
centrifuge at 16,000 g, 10 min

supernatant   pellet

4° C | add 20 g (NH$_4$)$_2$SO$_4$ / 100 ml;
centrifuge at 16,000 g, 10 min

supernatant   **pellet**

4° C | resuspend the pellet in buffer B;
dialyze against buffer B;
centrifuge at 10,000 g, 10 min;
chromatograph on

DEAE-column

4° C | elute with 250 ml linear gradient 10-380 mM NaCl;
dialyze overnight against buffer C;
chromatograph on

CM-52-cellulose column

4° C | elute with 100 ml linear gradient
100-500 mM NaCl in buffer C;

VINCULIN

VITAMIN D BINDING PROTEIN (DBP)

A single 55 kda polypeptide functioning as Vitamin D carrier and actin sequestering protein in the extracellular space of vertebrate tissues.

Source: human plasma

Equipment:
- Affi-Gel-15-G-actin affinity column (10 ml glass pipettes)
- hydroxylapatide column

Chemicals: EDTA, HEPES, Affi-Gel-15, Tris-HCl, $CaCl_2$, ATP, NaN_3, DTT, NaCl, guanidinium chloride, Na/PO_4

Have ready: **depolymerizing buffer:** 0.01 M Tris-HCl, 0.2 mM $CaCl_2$, 0.5 mM ATP, 1 mM NaN_3, 0.2 mM DTT, pH 8.0

3 M guanidinium chloride

0.01 M sodium phosphate, pH 7.0

0.2 M sodium phosphate, pH 7.0

Affi-Gel-15 / G-actin affinity column:
wash Affi-Gel-15 with cold deionized H_2O (200 ml); dialyze G-actin against 0.01 M HEPES, pH 7.5 at 4°C overnight; concentrate on Amicon PM-10 membrane to 5 mg/ml; add 1 ml of gel to 1 ml 5 mg/ml G-actin; incubate with rotation overnight at 4°C; pour column

Reference: Haddad J G Kowalski M A and Sanger J W (1984) Biochem. J. 218, 805-810
Gen-Bank: M60197 (human), M60206 (rat)

human plasma

collect into 2 mM EDTA and apply directly to
10 ml glass pipette columns of

Affi-gel-15-G-actin

4° C

rinse with depolymerizing buffer until OD280 is zero;
rinse with 0.75 M NaCl in depolymerizing buffer;
elute with 4 column volumes 3 M
guanidinium-chloride;
dialyze eluate against 2 x 75 vol. 0.01 M Tris, pH 8.0
overnight;
dialyze against 0.01 M Na/PO_4, pH 7.0 and apply to

hydroxylapatide column

equilibrated in 0.01 M Na/PO_4, pH 7.0;
rinse with column buffer and elute by a linear
gradient of 2 x 150 ml 0.01 M and 0.2 M
Na/PO_4 at pH 7.0 (22 ml/h);
collect 3.7 ml fractions

VITAMIN D BINDING PROTEIN (DBP)

147

ZYXIN

A 82 kDa monomeric protein present in adherens junctions and focal contacts. A zinc binding metalloprotein with high proline content. Distributed along actin filament bundles.

Source: chicken gizzard smooth muscle

Equipment:
- Waring blender
- water bath
- DEAE-cellulose column (8.0 x 2.5 cm)
- phenyl-sepharose CL-4B column (10 x 0.7 cm)
- SDS-PAGE
- HPLC-hydroxylapatite column (BioRad) 0.5 ml

Chemicals: Tris-HCl, EGTA, NaOH, acetic acid, $MgCl_2$, $(NH_4)_2SO_4$, Tris-acetate, NaCl, ethylene glycol
protease inhibitors: PMSF
protease inhibitor mixture: 0.1 mM PMSF, 0.1
mM   benzamidine HCl, 1 ng/ml pepstatin A, 1 ng/ml 1.10 phenanthroline final conc.

Have ready: **buffer A:** 2 mM Tris-HCl, 1 mM EGTA, 0.7 mM PMSF, pH 8.8

2 M NaOH

0.5 M acetic acid

1 M $MgCl_2$

buffer B-10: 20 mM Tris-acetate, 10 mM NaCl, 0.1% 2-mercaptoethanol, 0.1 mM EDTA, pH 7.6

HAP-buffer: 1 mM potassium phosphate (mono-and dibasic), 10 mM NaCl, 0.1 mM EDTA, 0.1% 2-mercaptoethanol, pH 7.2

Reference: Crawford, A.W. and Beckerle, M.C. (1991) J Biol. Chem 266, 5847-5853

148

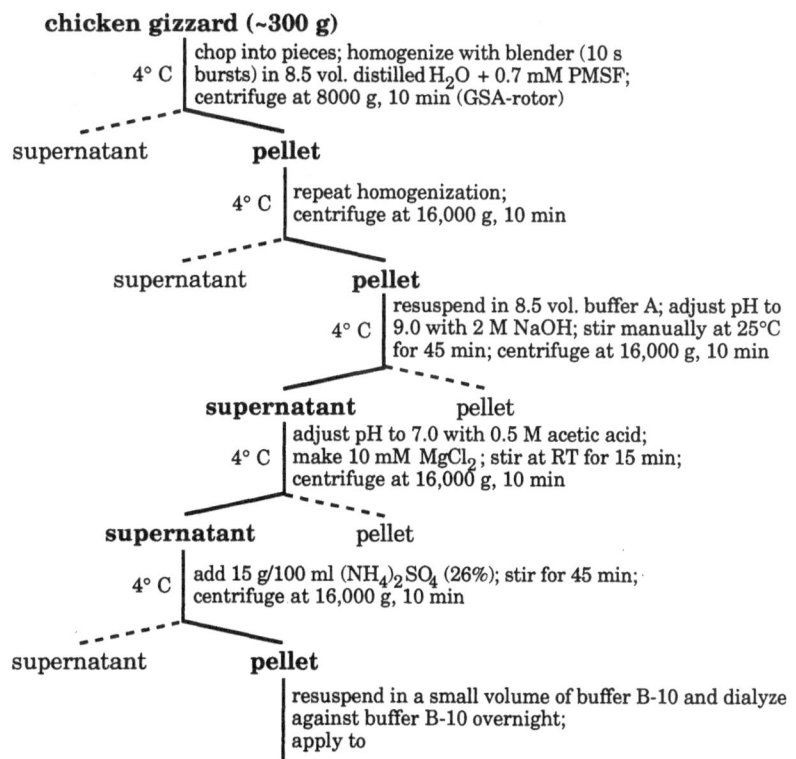

chicken gizzard (~300 g)

4° C | chop into pieces; homogenize with blender (10 s bursts) in 8.5 vol. distilled H_2O + 0.7 mM PMSF; centrifuge at 8000 g, 10 min (GSA-rotor)

supernatant   **pellet**

4° C | repeat homogenization; centrifuge at 16,000 g, 10 min

supernatant   **pellet**

4° C | resuspend in 8.5 vol. buffer A; adjust pH to 9.0 with 2 M NaOH; stir manually at 25°C for 45 min; centrifuge at 16,000 g, 10 min

supernatant   pellet

4° C | adjust pH to 7.0 with 0.5 M acetic acid; make 10 mM $MgCl_2$; stir at RT for 15 min; centrifuge at 16,000 g, 10 min

supernatant   pellet

4° C | add 15 g/100 ml $(NH_4)_2SO_4$ (26%); stir for 45 min; centrifuge at 16,000 g, 10 min

supernatant   **pellet**

resuspend in a small volume of buffer B-10 and dialyze against buffer B-10 overnight; apply to

DEAE-cellulose column

equilibrated in buffer B-10;
elute into plastic ! tubes containing protease inhibitor mixture by applying a linear gradient (120 ml) 0-150 mM NaCl in buffer B-10;
collect 1.25 ml fractions at 0.4 ml/min; pool and add $(NH_4)_2SO_4$ to 12.5%; apply to

phenyl-sepharose CL-4B column

equilibrated in buffer B-10 incl. 12.5% $(NH_4)_2SO_4$;
elute at 0.3 ml/min with 20 ml descending linear gradient of 12.5 - 0% $(NH_4)_2SO_4$ in buffer B-10 followed immediately by a 20 ml gradient of 0 - 50% ethylene glycol in buffer B-10;
check by SDS-PAGE; pool and add equal volume of HAP-buffer; apply to

HPLC-hydroxylapatide column

equilibrated with HAP-buffer;
elute with 15 ml linear gradient 0-75 mM potassium phosphate; collect 0.5 ml fractions at 0.5 ml/min

ZYXIN (~20 μg)

149

II Tubulin and Microtubule-Associated Proteins

BUTTONIN

A 75 kDa spherical molecule of ~9 nm in diameter, probably dimeric (~150 kDa) in its native form. A heat unstable, non-calmodulin binding MAP of sea urchin eggs.

Source:	sea urchin eggs
Equipment:	• water bath • centrifuge • BioGel A-1.5 (Bio Rad) gel filtration column (1 x 50 cm) • SDS-PAGE
Chemicals:	PIPES, EGTA, MgCl$_2$, glycerol, taxol, GTP, sucrose, NaCl **protease inhibitors:** 1 mM PMSF, 10 µg/ml leupeptin, 10 µU/ml aprotinin
Have ready:	**sea water** **PEM-buffer:** 0.1 M PIPES, 1 mM EGTA, 1 mM MgCl$_2$, pH 6.7
Reference:	Hirokawa, N. and Hisanaga, S. (1987) J. Cell Biol. 104, 1553-1561

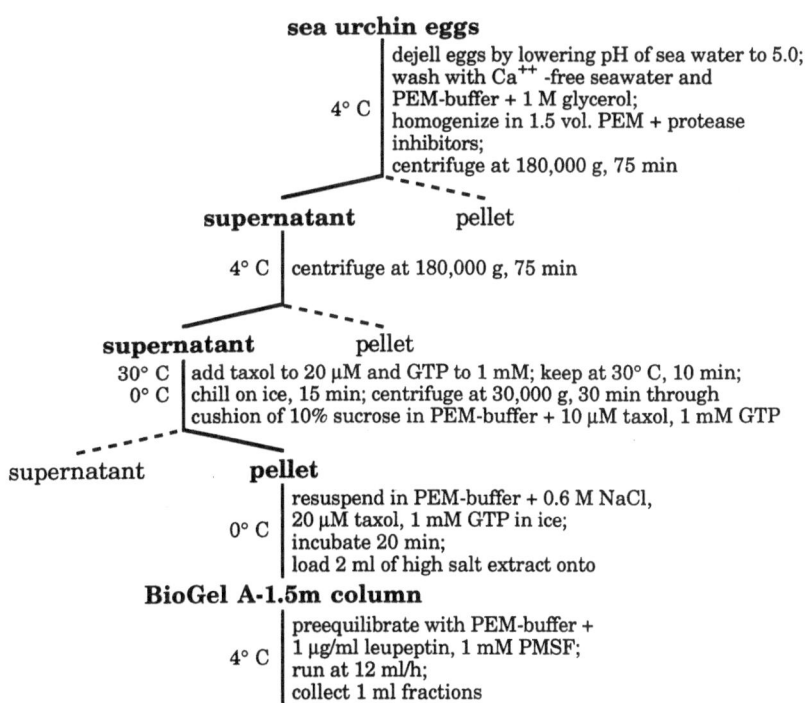

sea urchin eggs

4° C | dejell eggs by lowering pH of sea water to 5.0; wash with Ca^{++}-free seawater and PEM-buffer + 1 M glycerol; homogenize in 1.5 vol. PEM + protease inhibitors; centrifuge at 180,000 g, 75 min

supernatant   pellet

4° C | centrifuge at 180,000 g, 75 min

supernatant   pellet

30° C | add taxol to 20 µM and GTP to 1 mM; keep at 30° C, 10 min;
0° C | chill on ice, 15 min; centrifuge at 30,000 g, 30 min through cushion of 10% sucrose in PEM-buffer + 10 µM taxol, 1 mM GTP

supernatant   **pellet**

0° C | resuspend in PEM-buffer + 0.6 M NaCl, 20 µM taxol, 1 mM GTP in ice; incubate 20 min; load 2 ml of high salt extract onto

BioGel A-1.5m column

4° C | preequilibrate with PEM-buffer + 1 µg/ml leupeptin, 1 mM PMSF; run at 12 ml/h; collect 1 ml fractions

BUTTONIN

CLIP-170

A phosphorylation dependent microtubular (+)-end binding protein of 170 kDa. Essential as linker protein between endocytic vesicles and MTs.

Source:	Hela-cells
Equipment:	• Dounce homogenizer • centrifuge, Ti-70 rotor, TLA-100 rotor, SS-34 rotor (Beckman Instr.) • CNBr-activated sepharose 4B/IgG-affinity column (size variable)
Chemicals:	K-PIPES, MgSO$_4$, EGTA, DTT, Triton X-100, diethylamine, KCl, hexokinase, glucose **protease inhibitors:** PMSF, aprotinin, leupeptin, pepstatin
Have ready:	**PEM-buffer:** 0.1 M K-PIPES, 1 mM MgSO$_4$, 2 mM EGTA, pH 6.8 **homogenization solution:** 0.1 M K-PIPES, pH 6.8, 1 mM DTT, 1 mM PMSF, 1 µg/ml each aprotinin, leupeptin, pepstatin **phosphate buffered saline** (PBS)
Reference:	Rickard, J.E. and Kreis, T.E. (1991) J. Biol. Chem. 266, 17597-17605 Gen-Bank: M97501

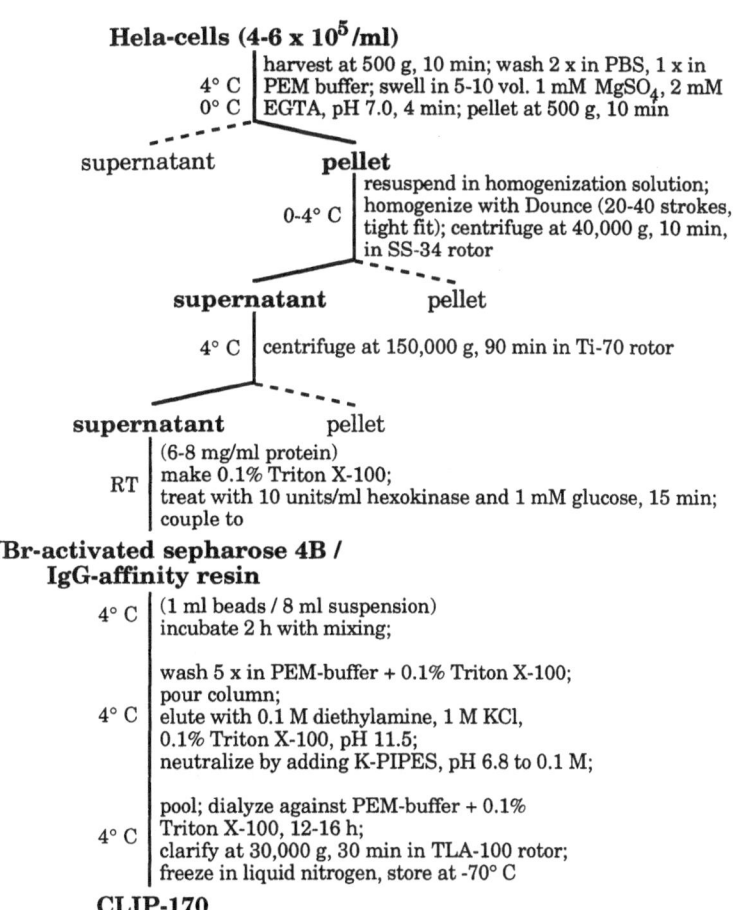

Hela-cells (4-6 x 10^5/ml)

4° C   harvest at 500 g, 10 min; wash 2 x in PBS, 1 x in
0° C   PEM buffer; swell in 5-10 vol. 1 mM MgSO$_4$, 2 mM
EGTA, pH 7.0, 4 min; pellet at 500 g, 10 min

supernatant   **pellet**

0-4° C   resuspend in homogenization solution;
homogenize with Dounce (20-40 strokes,
tight fit); centrifuge at 40,000 g, 10 min,
in SS-34 rotor

supernatant   pellet

4° C   centrifuge at 150,000 g, 90 min in Ti-70 rotor

supernatant   pellet

RT   (6-8 mg/ml protein)
make 0.1% Triton X-100;
treat with 10 units/ml hexokinase and 1 mM glucose, 15 min;
couple to

CNBr-activated sepharose 4B /
    IgG-affinity resin

4° C   (1 ml beads / 8 ml suspension)
incubate 2 h with mixing;

4° C   wash 5 x in PEM-buffer + 0.1% Triton X-100;
pour column;
elute with 0.1 M diethylamine, 1 M KCl,
0.1% Triton X-100, pH 11.5;
neutralize by adding K-PIPES, pH 6.8 to 0.1 M;

4° C   pool; dialyze against PEM-buffer + 0.1%
Triton X-100, 12-16 h;
clarify at 30,000 g, 30 min in TLA-100 rotor;
freeze in liquid nitrogen, store at -70° C

CLIP-170
(~15 µg / ml packed cells)

CYTOPLASMIC DYNEIN (MAP-1C)

A microtubular motor protein with 1,200 kDa molecular mass consisting of nine subunits of 2 x 410 kDa ATP-ase carrying heavy chains, 74 kDa and lower molecular mass subdomains. A motor protein responsible for retrograde organelle translocation and chromosome movement.

Source: calf brain

Equipment:
- teflon homogenizer
- centrifuge, GSA-rotor, SS-34 rotor
- polycarbonate tubes
- water bath
- centrifuge
- SW Ti-41 rotor

Chemicals: PIPES-NaOH, HEPES, $MgCl_2$, EDTA, DTT, taxol, Mg-GTP, Mg-ATP, Tris-HCl, $MgSO_4$, KCl, sucrose
protease inhibitors: PMSF, TAME, pepstatin A, leupeptin

Have ready: **extraction buffer:** 250 ml 0.05 M PIPES-NaOH, 0.05 M HEPES, 2 mM $MgCl_2$, 1 mM EDTA, 1 mM PMSF, 10 µg/ml leupeptin, 10 µg/ml tosyl-arginine methyl ester (TAME), 1 µg/ml pepstatin A, 1 mM DTT, pH 7.0

nucleotide extraction buffer: extraction buffer + 3 mM Mg-GTP (Sigma type II, 2 ml, 100 mM, make equimolar with $MgSO_4$, adjust pH to 7.0 on ice with NaOH)

Mg-ATP: Sigma type II, 1 ml, 100 mM, make equimolar with $MgSO_4$, adjust pH to 7.0 on ice with NaOH

mM **Tris-KCl buffer:** 20 mM Tris-HCl, 50 mM KCl, 5 $MgSO_4$, 0.5 mM EDTA, 1 mM DTT, pH 7.6

Reference: Paschal, B.M. et al. (1991) Methods Enzymol. 196, 181-191
Gen-Bank: X62160

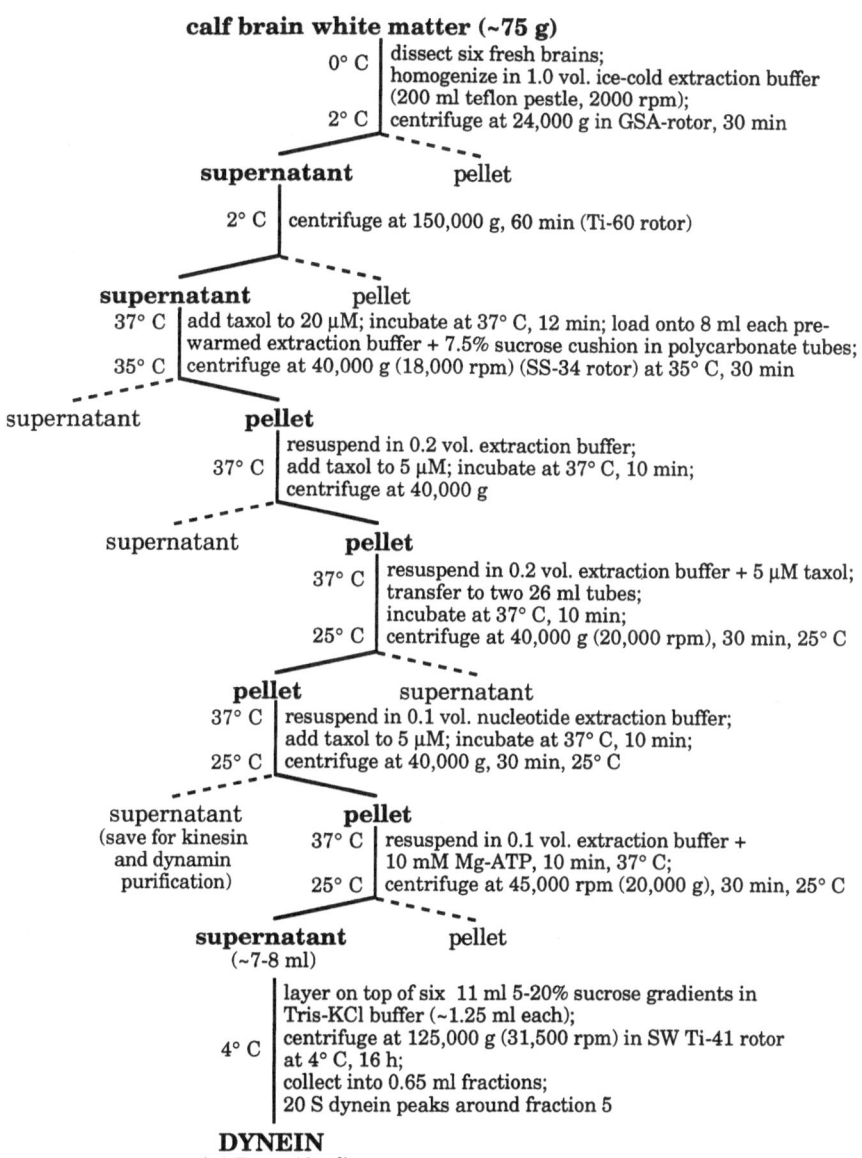

calf brain white matter (~75 g)

0° C | dissect six fresh brains;
      | homogenize in 1.0 vol. ice-cold extraction buffer
      | (200 ml teflon pestle, 2000 rpm);
2° C | centrifuge at 24,000 g in GSA-rotor, 30 min

supernatant   pellet

2° C | centrifuge at 150,000 g, 60 min (Ti-60 rotor)

supernatant   pellet

37° C | add taxol to 20 µM; incubate at 37° C, 12 min; load onto 8 ml each pre-
       | warmed extraction buffer + 7.5% sucrose cushion in polycarbonate tubes;
35° C | centrifuge at 40,000 g (18,000 rpm) (SS-34 rotor) at 35° C, 30 min

supernatant   **pellet**

       | resuspend in 0.2 vol. extraction buffer;
37° C | add taxol to 5 µM; incubate at 37° C, 10 min;
       | centrifuge at 40,000 g

supernatant   **pellet**

37° C | resuspend in 0.2 vol. extraction buffer + 5 µM taxol;
       | transfer to two 26 ml tubes;
       | incubate at 37° C, 10 min;
25° C | centrifuge at 40,000 g (20,000 rpm), 30 min, 25° C

pellet   supernatant

37° C | resuspend in 0.1 vol. nucleotide extraction buffer;
       | add taxol to 5 µM; incubate at 37° C, 10 min;
25° C | centrifuge at 40,000 g, 30 min, 25° C

supernatant   **pellet**
(save for kinesin   37° C | resuspend in 0.1 vol. extraction buffer +
and dynamin | 10 mM Mg-ATP, 10 min, 37° C;
purification)   25° C | centrifuge at 45,000 rpm (20,000 g), 30 min, 25° C

supernatant   pellet
(~7-8 ml)

       | layer on top of six 11 ml 5-20% sucrose gradients in
       | Tris-KCl buffer (~1.25 ml each);
4° C | centrifuge at 125,000 g (31,500 rpm) in SW Ti-41 rotor
       | at 4° C, 16 h;
       | collect into 0.65 ml fractions;
       | 20 S dynein peaks around fraction 5

DYNEIN
(~0.7 mg, 12 ml)

205 kDa MAP

A partially purified thermostable MAP protein from Drosophila with an apparent molecular weight of 205 kDa. Several isoforms.

Source: Drosophila melanogaster

Equipment:
- sonifier (Branson)
- Potter with teflon pestle
- centrifuge, SW 60-Ti, 65-Ti rotors (Beckman Instr.)
- boiling water bath

Chemicals: PIPES, $MgCl_2$, EGTA, GTP, DTT, NaCl, sucrose, taxol
protease inhibitors: PMSF, leupeptin, pepstatin A, aprotinin, TAME

Have ready: **MT-assembly buffer:** 0.1 M PIPES, 1 mM $MgCl_2$, 2 mM EGTA, 2 mM DTT, 0.1 mM GTP, 1 mM PMSF, 1 µg/ml leupeptin, 1 µg/ml pepstatin A, 2 µg/ml aprotinin, 2 mg/ml tosyl-arginyl-methyl ester (TAME), pH 6.6

Reference: Goldstein, L.S.B. et al. (1986) J. Cell Biol. 102, 2076-2087

Drosophila "Schneider S-2 cells"

(1 l, ~5 g packed cells)

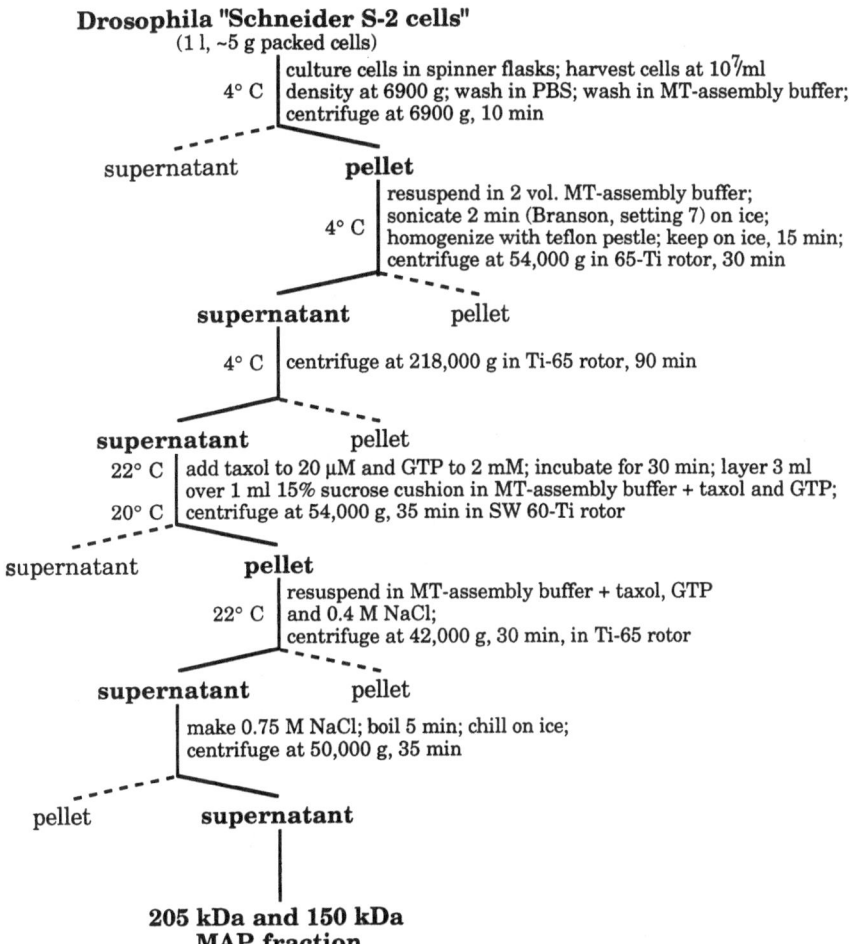

4° C | culture cells in spinner flasks; harvest cells at 10^7/ml density at 6900 g; wash in PBS; wash in MT-assembly buffer; centrifuge at 6900 g, 10 min

supernatant   **pellet**

4° C | resuspend in 2 vol. MT-assembly buffer; sonicate 2 min (Branson, setting 7) on ice; homogenize with teflon pestle; keep on ice, 15 min; centrifuge at 54,000 g in 65-Ti rotor, 30 min

supernatant   pellet

4° C | centrifuge at 218,000 g in Ti-65 rotor, 90 min

supernatant   pellet

22° C | add taxol to 20 µM and GTP to 2 mM; incubate for 30 min; layer 3 ml over 1 ml 15% sucrose cushion in MT-assembly buffer + taxol and GTP;
20° C | centrifuge at 54,000 g, 35 min in SW 60-Ti rotor

supernatant   **pellet**

22° C | resuspend in MT-assembly buffer + taxol, GTP and 0.4 M NaCl; centrifuge at 42,000 g, 30 min, in Ti-65 rotor

supernatant   pellet

make 0.75 M NaCl; boil 5 min; chill on ice; centrifuge at 50,000 g, 35 min

pellet   **supernatant**

205 kDa and 150 kDa MAP fraction

MAP-1A / MAP-1B

MAP-1A, a rod shaped molecule consisting of a ~350 kDa heavy chain and three light chains of 30 kDa, 28 kDa and 18 kDa. Builds armlike projections and stabilizes MTs in mature neuronal tissue.
MAP-1B represents similar molecules of 10 x 200 nm with a putative role in neuronal maturation.

Source:	rat brain
Equipment:	• water bath • phosphocellulose P-11 column (0.5 x 5 cm) • Bio-Gel A-15 column (1.2 x 50 cm) • SDS-PAGE (5-12% gradient gels)
Chemicals:	MES, Mg(CH$_3$COO)$_2$, DTT, EGTA, Poly-(L-aspartic acid) 47 kDa PLAA, taxol, KCl **protease inhibitors:** PMSF, pepstatin
Have ready:	**buffer B:** 100 mM MES, 0.5 mM Mg(CH$_3$COO)$_2$, 1 mM DTT, 1 mM EGTA, 0.5 mM PMSF, pH 6.5 + 10 µg/ml pepstatin
	buffer A: 10 mM MES, 0.5 mM Mg(CH$_3$COO)$_2$, 0.1 mM EGTA, 1 mM DTT, pH 6.5 + 0.5 mM PMSF
Reference:	Fujii, T. et al. (1990) Analyt. Biochem. 184, 268-273 * Shiomura, Y. and Hirokawa, N. (1987) J. Neurosci. 7, 1461-1469 * Sato-Yoshitake et al. (1989) Neuron 3, 229-238

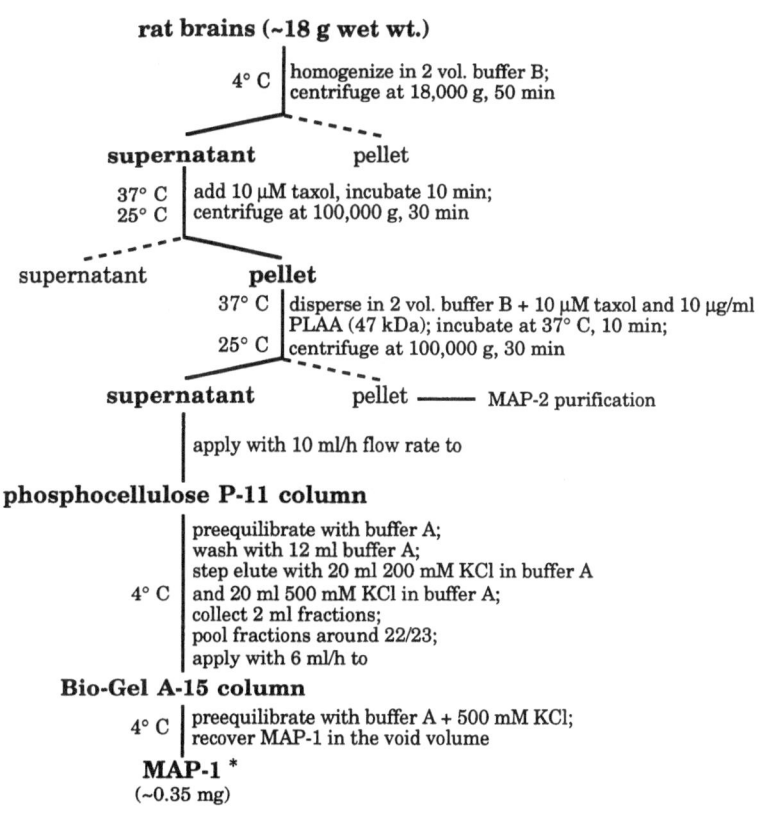

rat brains (~18 g wet wt.)

4° C | homogenize in 2 vol. buffer B;
centrifuge at 18,000 g, 50 min

supernatant   pellet

37° C | add 10 μM taxol, incubate 10 min;
25° C | centrifuge at 100,000 g, 30 min

supernatant   **pellet**

37° C | disperse in 2 vol. buffer B + 10 μM taxol and 10 μg/ml
PLAA (47 kDa); incubate at 37° C, 10 min;
25° C | centrifuge at 100,000 g, 30 min

supernatant   pellet —— MAP-2 purification

apply with 10 ml/h flow rate to

phosphocellulose P-11 column

preequilibrate with buffer A;
wash with 12 ml buffer A;
step elute with 20 ml 200 mM KCl in buffer A
4° C | and 20 ml 500 mM KCl in buffer A;
collect 2 ml fractions;
pool fractions around 22/23;
apply with 6 ml/h to

Bio-Gel A-15 column

4° C | preequilibrate with buffer A + 500 mM KCl;
recover MAP-1 in the void volume

MAP-1 *
(~0.35 mg)

* (The only way to purify MAP-1A from MAP-1B is to apply an
antibody affinity column for either protein. See reference)

MAP-2

A neuronal specific high molecular weight MT-binding phosphoprotein consisting of 280 kDa (MAP-2a), 270 kDa (MAP-2b) and 70 kDa (MAP-2c). MAP-2 is also an actin and neurofilament binding protein. Precise function yet unknown.

Source: calf brain

Equipment:
- centrifuge, GSA-rotor, SS-34 rotor
- polypropylene tubes
- Waring blender
- teflon/glass homogenizer
- water bath
- DEAE-sephadex A-50 column (Pharmacia) (4 ml)
- Bio-Gel A-15m (Bio Rad) gel-filtration column (1.5 x 25 or 0.9 x 55 cm)

Chemicals: PIPES-NaOH, EGTA, $MgSO_4$, 2-mercaptoethanol, GTP, ATP, sucrose, NaCl

Have ready: **PEM-buffer:** (1.1 l) 0.1 M PIPES-NaOH, 1 mM EGTA, 1 mM $MgSO_4$, pH 6.6

GTP, Sigma Type II, 150 mg
ATP, Sigma Type II, 1.1 g

sucrose underlayer solution: 90 ml PEM-buffer, 10% sucrose, 1 mM GTP (make fresh !)

dissociation buffer: 1 ml PEM-buffer + 0.25 M NaCl, 0.1 mM GTP

sephadex column buffer: 100 ml PEM-buffer + 0.25 M NaCl + 0.1 mM GTP

sephadex elution buffer: 10 ml PEM-buffer + 0.5 M NaCl + 0.1 mM GTP

dialysis buffer: 250 ml PEM-buffer + 0.1 mM GTP

Reference: Vallee, R.B. (1986) Methods Enzymol. 134, 89-115
Gen-Bank: M21041 (mouse), X51842 (rat), M25668 (human)

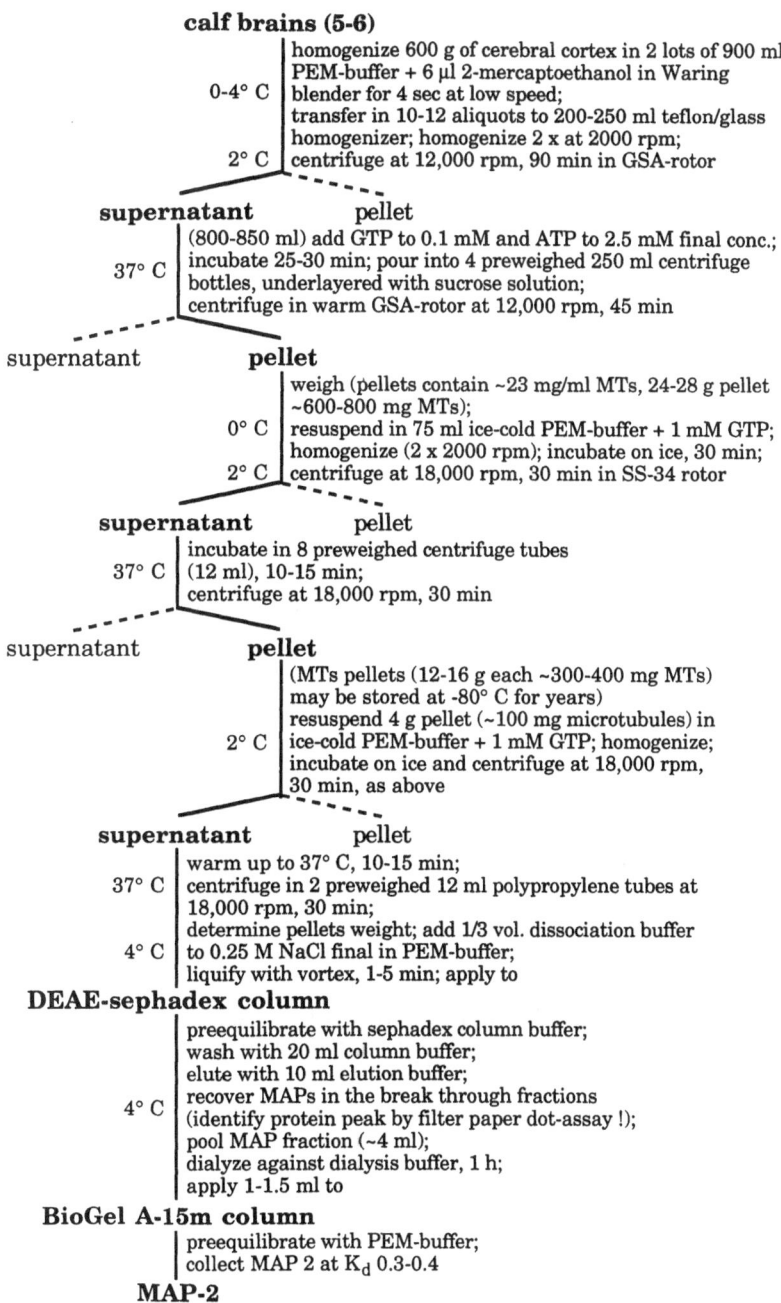

calf brains (5-6)

0-4° C | homogenize 600 g of cerebral cortex in 2 lots of 900 ml PEM-buffer + 6 µl 2-mercaptoethanol in Waring blender for 4 sec at low speed; transfer in 10-12 aliquots to 200-250 ml teflon/glass homogenizer; homogenize 2 x at 2000 rpm;
2° C | centrifuge at 12,000 rpm, 90 min in GSA-rotor

supernatant   pellet

37° C | (800-850 ml) add GTP to 0.1 mM and ATP to 2.5 mM final conc.; incubate 25-30 min; pour into 4 preweighed 250 ml centrifuge bottles, underlayered with sucrose solution; centrifuge in warm GSA-rotor at 12,000 rpm, 45 min

supernatant   **pellet**

0° C | weigh (pellets contain ~23 mg/ml MTs, 24-28 g pellet ~600-800 mg MTs); resuspend in 75 ml ice-cold PEM-buffer + 1 mM GTP;
2° C | homogenize (2 x 2000 rpm); incubate on ice, 30 min; centrifuge at 18,000 rpm, 30 min in SS-34 rotor

supernatant   pellet

37° C | incubate in 8 preweighed centrifuge tubes (12 ml), 10-15 min; centrifuge at 18,000 rpm, 30 min

supernatant   **pellet**

2° C | (MTs pellets (12-16 g each ~300-400 mg MTs) may be stored at -80° C for years) resuspend 4 g pellet (~100 mg microtubules) in ice-cold PEM-buffer + 1 mM GTP; homogenize; incubate on ice and centrifuge at 18,000 rpm, 30 min, as above

supernatant   pellet

37° C | warm up to 37° C, 10-15 min; centrifuge in 2 preweighed 12 ml polypropylene tubes at 18,000 rpm, 30 min;
4° C | determine pellets weight; add 1/3 vol. dissociation buffer to 0.25 M NaCl final in PEM-buffer; liquify with vortex, 1-5 min; apply to

DEAE-sephadex column

4° C | preequilibrate with sephadex column buffer; wash with 20 ml column buffer; elute with 10 ml elution buffer; recover MAPs in the break through fractions (identify protein peak by filter paper dot-assay !); pool MAP fraction (~4 ml); dialyze against dialysis buffer, 1 h; apply 1-1.5 ml to

BioGel A-15m column

preequilibrate with PEM-buffer; collect MAP 2 at K_d 0.3-0.4

MAP-2
(0.5 mg/ml, peak fraction)

MAP-4 (190 kDa)

A family of ~200 kDa proteins, consisting of long flexible rods of 100 nm contour length. A phosphorylation-inhibited promotor of MT-assembly, wide spread, but in brain restricted to glial cells.

Source:	bovine adrenal glands
Equipment:	• centrifuge
	• Waring blender
	• boiling water bath
	• DEAE-cellulose column (DE-52), Whatman (1.3 x 22 cm)
	• Butyl-Toyopearl 650 l (hydrophobic chromatography), Toyo Soda Co Ltd. Tokyo (1 x 8 cm)
Chemicals:	PIPES, EGTA, $MgCl_2$, NaCl, 2-mercaptoethanol, MES, $(NH_4)_2SO_4$, GTP, sucrose
	protease inhibitors: PMSF, leupeptin, pepstatin
Have ready:	**extraction solution:** 0.1 M PIPES, 2 mM EGTA, 1 mM $MgCl_2$, pH 6.8 + 1 mM PMSF, 10 µg/ml leupeptin, 10 µg/ml pepstatin
	MEM-buffer: 20 mM MES, 1 mM EGTA, 0.5 mM $MgCl_2$, pH 6.8 + 0.2 mM PMSF, 1 µg/ml leupeptin, 1 µg/ml pepstatin
	dialyzing buffer: MEM-buffer + 1 M $(NH_4)_2SO_4$
	PEM-buffer: 0.1 M PIPES, 2 mM EGTA, 1 mM $MgCl_2$, pH 6.8
	two-cycle purified brain tubulin (5 mg/ml) in PEM-buffer
	sucrose cushion: 10% (wt/vol) sucrose, 0.5 mM GTP, 0.2 mM PMSF, 1 µg/ml leupeptin, 1 µg/ml pepstatin in PEM-buffer
Reference:	Murofushi, H. et al. (1986) J. Cell Biol. 103, 1911-1919
	Gen-Bank: J05557 (bovine), M64571 (human), M72414 (mouse)

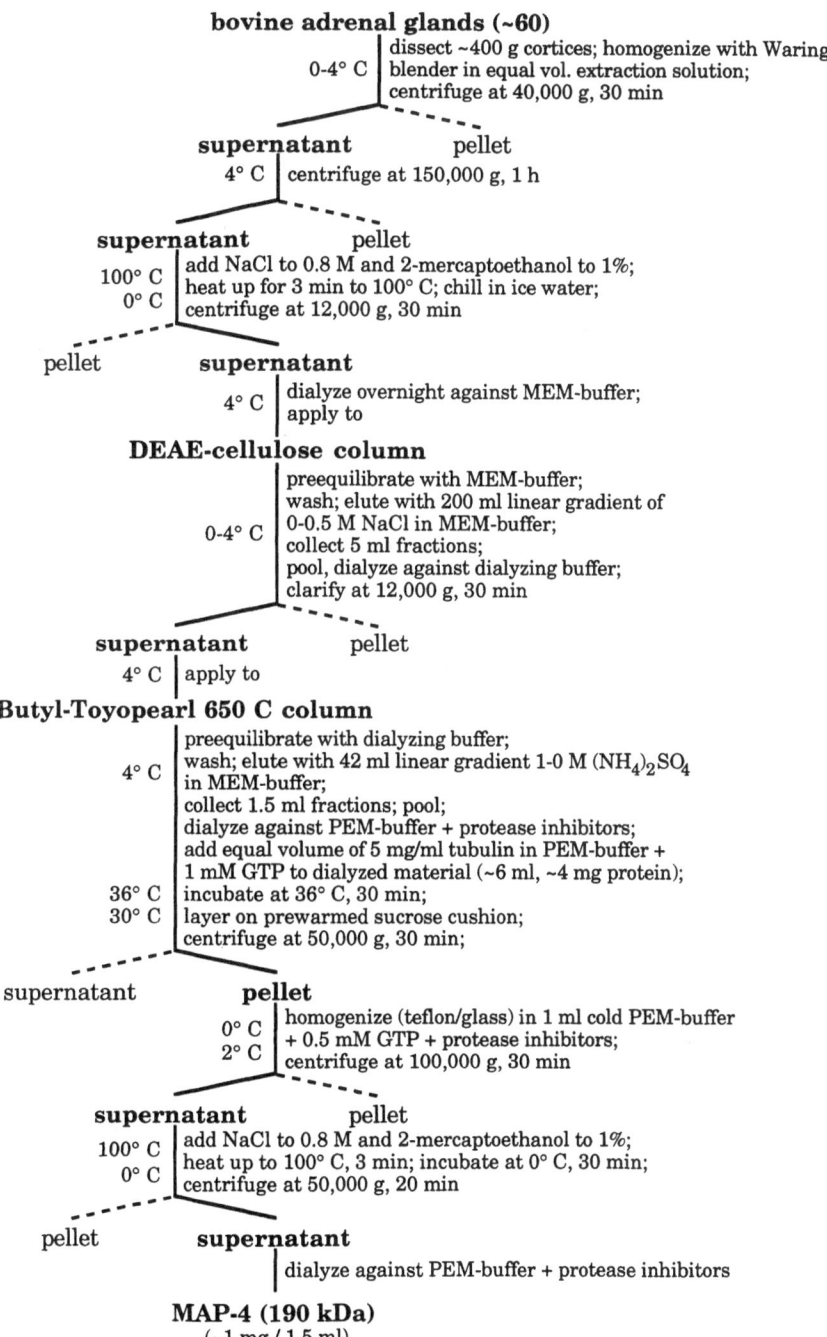

bovine adrenal glands (~60)

0-4° C | dissect ~400 g cortices; homogenize with Waring blender in equal vol. extraction solution; centrifuge at 40,000 g, 30 min

supernatant   pellet

4° C | centrifuge at 150,000 g, 1 h

supernatant   pellet

100° C | add NaCl to 0.8 M and 2-mercaptoethanol to 1%;
0° C | heat up for 3 min to 100° C; chill in ice water; centrifuge at 12,000 g, 30 min

pellet

supernatant

4° C | dialyze overnight against MEM-buffer; apply to

DEAE-cellulose column

0-4° C | preequilibrate with MEM-buffer; wash; elute with 200 ml linear gradient of 0-0.5 M NaCl in MEM-buffer; collect 5 ml fractions; pool, dialyze against dialyzing buffer; clarify at 12,000 g, 30 min

supernatant   pellet

4° C | apply to

Butyl-Toyopearl 650 C column

4° C | preequilibrate with dialyzing buffer; wash; elute with 42 ml linear gradient 1-0 M $(NH_4)_2SO_4$ in MEM-buffer; collect 1.5 ml fractions; pool; dialyze against PEM-buffer + protease inhibitors; add equal volume of 5 mg/ml tubulin in PEM-buffer + 1 mM GTP to dialyzed material (~6 ml, ~4 mg protein);
36° C | incubate at 36° C, 30 min;
30° C | layer on prewarmed sucrose cushion; centrifuge at 50,000 g, 30 min;

supernatant   **pellet**

0° C | homogenize (teflon/glass) in 1 ml cold PEM-buffer
2° C | + 0.5 mM GTP + protease inhibitors; centrifuge at 100,000 g, 30 min

supernatant   pellet

100° C | add NaCl to 0.8 M and 2-mercaptoethanol to 1%;
0° C | heat up to 100° C, 3 min; incubate at 0° C, 30 min; centrifuge at 50,000 g, 20 min

pellet   **supernatant**

| dialyze against PEM-buffer + protease inhibitors

MAP-4 (190 kDa)
(~1 mg / 1.5 ml)

MAP-4 (210 kDa)

A 210 kDa protein member of the MAP-4 family present in a variety of cells and tissues. A 100 nm long flexible rod, phosphorylatable by protein kinase C, cdc kinase and MAP-2 kinase, probably regulating MT-assembly during the cell cycle.

Source:	Hela-cells
Equipment:	• sonifier • centrifuge, SS-34 rotor, SW Ti-60 rotor • vacuum concentration equipment • sucrose gradient
Chemicals:	NaCl, KH_2PO_4, Na_2HPO_4, EGTA, PIPES, GTP, DTT, sucrose, KCl
Have ready:	**PBS (phosphate buffered saline):** 0.2 g/l KCl, 0.2 g /l KH_2PO_4, 8.0 g/l NaCl, 2.16 g/l Na_2HPO_4
	MT-assembly buffer: 0.1 M PIPES, 1 mM EGTA, 0.1 mM GTP, pH 6.9 + 1 mM DTT
Reference:	Bulinski, J.C. (1986) Methods Enzymol. 134, 147-156

166

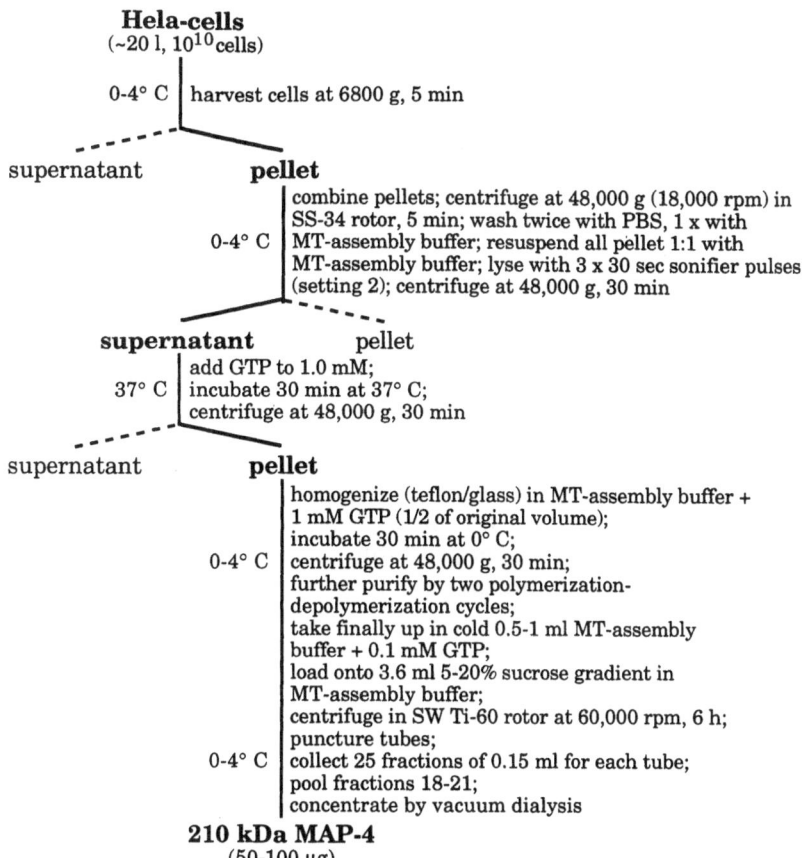

Hela-cells
(~20 l, 10^{10} cells)

0-4° C | harvest cells at 6800 g, 5 min

supernatant   **pellet**

0-4° C | combine pellets; centrifuge at 48,000 g (18,000 rpm) in SS-34 rotor, 5 min; wash twice with PBS, 1 x with MT-assembly buffer; resuspend all pellet 1:1 with MT-assembly buffer; lyse with 3 x 30 sec sonifier pulses (setting 2); centrifuge at 48,000 g, 30 min

supernatant   pellet

37° C | add GTP to 1.0 mM; incubate 30 min at 37° C; centrifuge at 48,000 g, 30 min

supernatant   **pellet**

0-4° C | homogenize (teflon/glass) in MT-assembly buffer + 1 mM GTP (1/2 of original volume); incubate 30 min at 0° C; centrifuge at 48,000 g, 30 min; further purify by two polymerization-depolymerization cycles; take finally up in cold 0.5-1 ml MT-assembly buffer + 0.1 mM GTP; load onto 3.6 ml 5-20% sucrose gradient in MT-assembly buffer; centrifuge in SW Ti-60 rotor at 60,000 rpm, 6 h; puncture tubes;

0-4° C | collect 25 fractions of 0.15 ml for each tube; pool fractions 18-21; concentrate by vacuum dialysis

210 kDa MAP-4
(50-100 µg)

MARPs

Heat stable high molecular weight (~320 kDa) MT-associated proteins in trypanosomatids likely to be involved in stabilizing MTs and linking them to the cell membrane.

Source: Trypanosoma brucei

Equipment:
- sonifier
- ice/water bath
- HB4-swingant rotor (Sorvall), SS-34 rotor
- FPLC-superose-12 gel filtration column

(Pharmacia)

Chemicals: MOPS, NaCl, $MgCl_2$, EGTA, Triton X-100, KCl, KH_2PO_4, Na_2HPO_4, $(NH_4)_2SO_4$, KCl
protease inhibitors: PMSF, leupeptin, chymostatin, pepstatin

Have ready: **MARP-buffer:** 100 mM MOPS (morpholino propane sulfonic acid), 50 mM NaCl, 5 mM $MgCl_2$, 1 mM EGTA, pH 6.9 + leupeptin, chymostatin, pepstatin at 5 µg/ml + 0.2 mM PMSF

PBS: 137 mM NaCl, 2.6 mM KCl, 8 mM Na_2HPO_4, 1.5 mM KH_2PO_4, pH 7.2

cold saturated $(NH_4)_2SO_4$

MME-buffer: 100 mM MOPS, 2 mM $MgCl_2$, 1 mM EGTA, pH 6.9

Reference: Hemphill, A. et al. (1992) J. Cell Biol. 117, 95-103
Gen-Bank: M20569

trypanosomes
(2×500 ml, 5×10^6-1×10^7 cells/ml)

$4°$ C | harvest at 1000 g, 10 min

supernatant   **pellet**

0-4° C | resuspend in 30 ml MARP-buffer; incubate on ice, 10 min;
add Triton X-100 to 0.5%; vortex; incubate on ice, 5 min;
centrifuge at 6000 g, 5 min (SS-34 rotor)

supernatant   **pellet**

4° C | resuspend in 30 ml MARP buffer +
0.1% Triton X-100;
centrifuge at 6000 g, 5 min

pellet   supernatant

0° C | add 12 ml PBS containing 1.0 M NaCl + protease inhibitors;
sonicate 4 x 5 sec in ice water; keep on ice, 30 min;
4° C | vortex in between; centrifuge at 10,000 g, 20 min

supernatant   pellet

100° C | boil, 5 min; chill on ice, 10 min;
add 0.1 mM PMSF;
4° C | centrifuge at 100,000 g, 20 min

pellet   **supernatant**

4° C | slowly add 1/10 vol. cold saturated $(NH_4)_2SO_4$ to 10%;
sediment at 10,000 g, 10 min, in HB4-rotor

supernatant   **pellet**

lyse in 500 μl MARP-buffer +
6 M urea, 0.1 mM PMSF;
load onto

FPLC-superose-12 gel filtration column

4° C | equilibrate and elute with the same buffer;
pool MARP fractions;
dialyze against MME-buffer with descending
urea conc. (4-2-1-0.5-0.25 M), 1 h each;
dialyze against MME-buffer +
0.1 mM PMSF (3 x 90 min)

MARPs
(store at -70° C in liquid nitrogen)

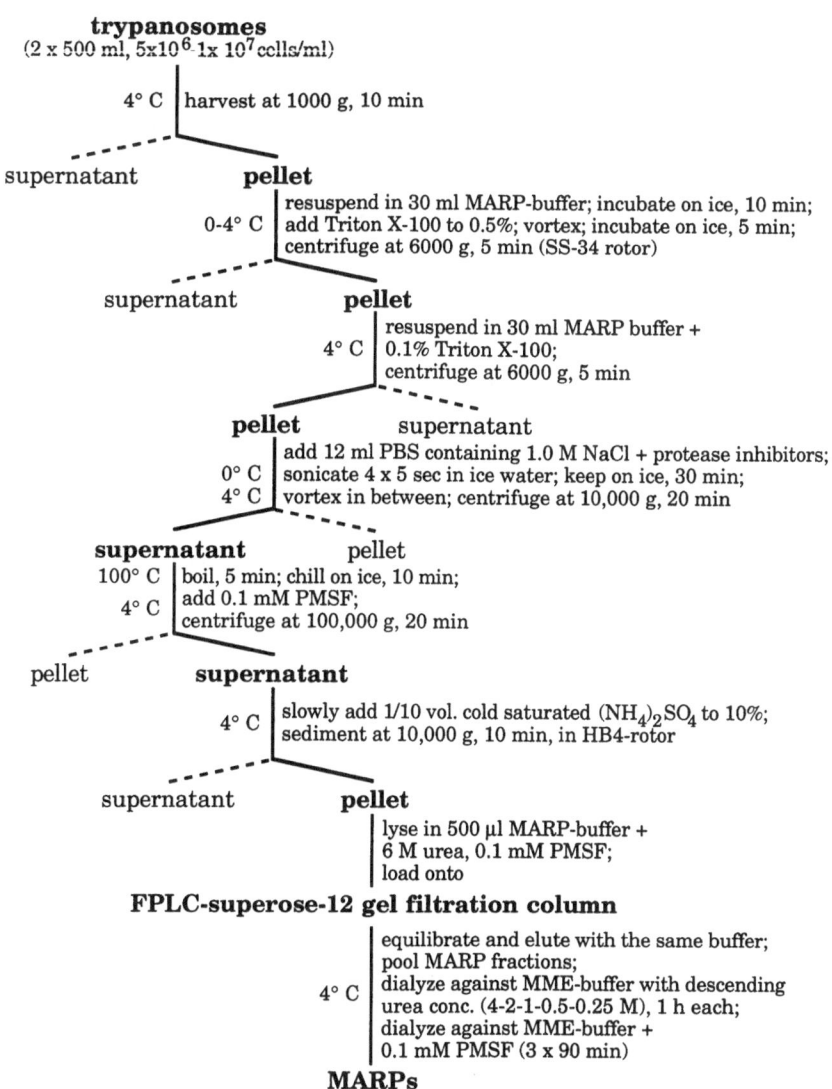

169

STOPs

STOPs, (stable tubulin only polypeptides), globular protein with an apparent MW of 145 kDa, responsible for cold stable MTs. Ca^{++}-calmodulin regulated, widespread.

Source:	rat brain
Equipment:	• Potter-Elvehjem (glass/teflon, loose fitting) • centrifuge, Ti-70.1 rotor (Beckman Instr.) • DEAE-ion exchange column (4.5 ml), DE-52 (Whatman) • calmodulin-affinity column (Bio Rad) (1.5 ml) • SDS-gels (10%)
Chemicals:	MES, EGTA, CaCl$_2$, MgSO$_4$, NaN$_3$, podophyllotoxin (PLN, Aldrich), sucrose, Triton X-100
Have ready:	**MME-buffer:** 100 mM MES, 1 mM MgSO$_4$, 1 mM EGTA, pH 6.75 + 0.02% NaN$_3$
Reference:	Margolis, R.L. et al. (1986) Methods Enzymol. 134, 160-170

8 rat brains

0° C | homogenize in 14 ml MME-buffer with Potter-Elvehjem (glass/teflon), 2400 rpm, three strokes; add 2 mM CaCl$_2$ during homogenization;
4° C | immediately add 4 mM EGTA after homogenization; centrifuge at 200,000 g, 30 min in Ti-70.1 rotor

supernatant   pellet

30° C | incubate for 60 min at 30° C; cool down to 0° C, 10 min;
0° C, RT | add 25 μM PLN; rewarm and layer onto 50% sucrose cushion in
25° C | MME-buffer; centrifuge at 200,000 g, 2 h

supernatant   **pellet**

0° C | resuspend in 0.75 ml MME-buffer; incubate with Ca^{++} for 20 min; shear through 26-gauge needle;
5° C | centrifuge at 40,000 g, 30 min

supernatant   pellet

| apply to

DEAE-ion exchange column

| (do not preequilibrate with MME !)
4° C | collect flow through fraction;
| make 1 mM free CaCl$_2$;
| apply 0.5 ml aliquots every 10 min to

calmodulin affinity column

| wash with Ca^{++} -MME-buffer + 0.1% Triton X-100;
| wash 2 x with 6 ml MME-buffer + 0.3 M KCl, 1.0 mM free CaCl$_2$;
| elute with MME-buffer + 0.1 M KCl, 0.1% Triton X-100;
| collect 0.5 ml fractions;
| pool peak (~2.5 ml)

STOP
(~200 μg)

SYNCOLIN

A 280 kDa MT-binding protein distinct from MAP-2, globular structure of ~13 nm diameter, induces bundling of microtubules.

Source:	chicken erythrocytes
Equipment:	• Branson microtip sonifier • Dounce homogenizer • water bath • sucrose gradient • centrifuge, SW-28 rotor • hydroxylapatite (Bio Rad) column (0.5 x 5 cm) • DEAE-sepharose CL-6B (Pharmacia) column (0.5 x 5 cm) • SDS-PAGE
Chemicals:	MES, $MgCl_2$, EGTA, sucrose, GTP, DTT, taxol, PIPES, Na_2HPO_4, NaCl **protease inhibitors:** PMSF
Have ready:	**buffer A:** 100 mM MES, 0.5 mM $MgCl_2$, 1 mM EGTA, 0.52 M sucrose, pH 7.0 + 1 mM PMSF, 1 mM DTT, 1 mM GTP **buffer B:** 100 mM PIPES, 1 mM EGTA, 1 mM $MgCl_2$, pH 6.6 + 1 mM GTP, 10 µM taxol **buffer C:** 100 mM MES, 1 mM EGTA, 1 mM $MgCl_2$, pH 6.7 **buffer D:** 50 mM Na_2HPO_4, 0.5 mM $MgCl_2$, pH 6.7 + 0.1 mM PMSF **storage buffer:** 25 mM MES, 1 mM EGTA, 1 mM $MgCl_2$, pH 6.7 + 0.1 mM PMSF
Reference:	Feick, P., Foisner, R. and Wiche, G. (1991) J. Cell Biol. 112, 686-699

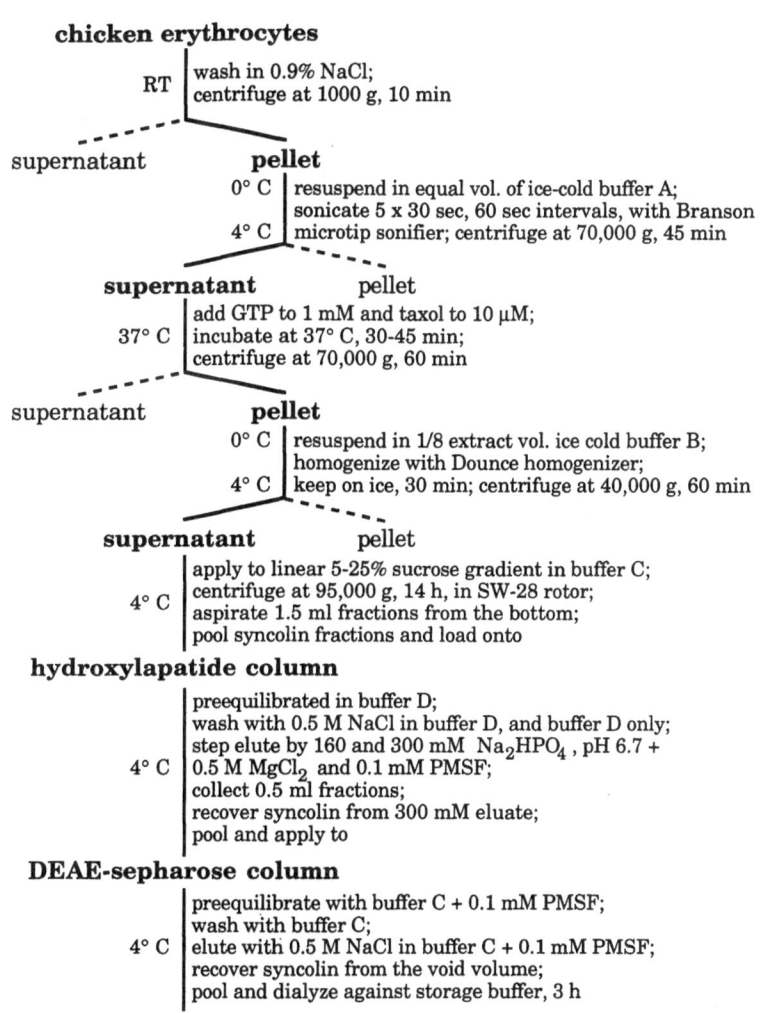

chicken erythrocytes

RT | wash in 0.9% NaCl;
centrifuge at 1000 g, 10 min

supernatant   **pellet**

0° C | resuspend in equal vol. of ice-cold buffer A;
sonicate 5 x 30 sec, 60 sec intervals, with Branson
4° C | microtip sonifier; centrifuge at 70,000 g, 45 min

supernatant   pellet

37° C | add GTP to 1 mM and taxol to 10 µM;
incubate at 37° C, 30-45 min;
centrifuge at 70,000 g, 60 min

supernatant   **pellet**

0° C | resuspend in 1/8 extract vol. ice cold buffer B;
homogenize with Dounce homogenizer;
4° C | keep on ice, 30 min; centrifuge at 40,000 g, 60 min

supernatant   pellet

4° C | apply to linear 5-25% sucrose gradient in buffer C;
centrifuge at 95,000 g, 14 h, in SW-28 rotor;
aspirate 1.5 ml fractions from the bottom;
pool syncolin fractions and load onto

hydroxylapatide column

4° C | preequilibrated in buffer D;
wash with 0.5 M NaCl in buffer D, and buffer D only;
step elute by 160 and 300 mM Na_2HPO_4 , pH 6.7 +
0.5 M $MgCl_2$ and 0.1 mM PMSF;
collect 0.5 ml fractions;
recover syncolin from 300 mM eluate;
pool and apply to

DEAE-sepharose column

4° C | preequilibrate with buffer C + 0.1 mM PMSF;
wash with buffer C;
elute with 0.5 M NaCl in buffer C + 0.1 mM PMSF;
recover syncolin from the void volume;
pool and dialyze against storage buffer, 3 h

SYNCOLIN

TAU

An isoform family of 35-65 kDa MAPs. A phosphorylatable protein stimulating nucleation and elongation of MTs, wide spread in vertebrates.

Source: bovine brain

Equipment:
- boiling water bath
- centrifuge, polycarbonate tubes
- Dounce homogenizer
- sepharose CL-4B column (Pharmacia) (1.5 x 86 cm)
- SDS-PAGE

Chemicals: MES-KOH, $MgCl_2$, EGTA, EDTA, GTP, 2-mercaptoethanol, NaCl, DTT, $(NH_4)_2SO_4$

protease inhibitors: 10 µM benzamidine-HCl, 1 mM PMSF, 1 µg/ml phenanthroline, 10 µg/ml aprotinin, 10 µg/ml leupeptin, 10 µg/ml pepstatin A

Have ready: **PB-buffer:** 100 mM MES-KOH, 0.5 mM $MgCl_2$, 2.0 mM EGTA, 0.1 mM EDTA, pH 6.4 + 1.0 mM 2-mercaptoethanol, 1 mM GTP + 1 x protease inhibitors

column buffers are supplemented with 0.1 x protease inhibitors

Reference: Drubin, D. and Kirschner, M. (1986) Methods Enzymol. 134, 156-160
Gen-Bank: J03778 (human),
M18775, M18776 (mouse)

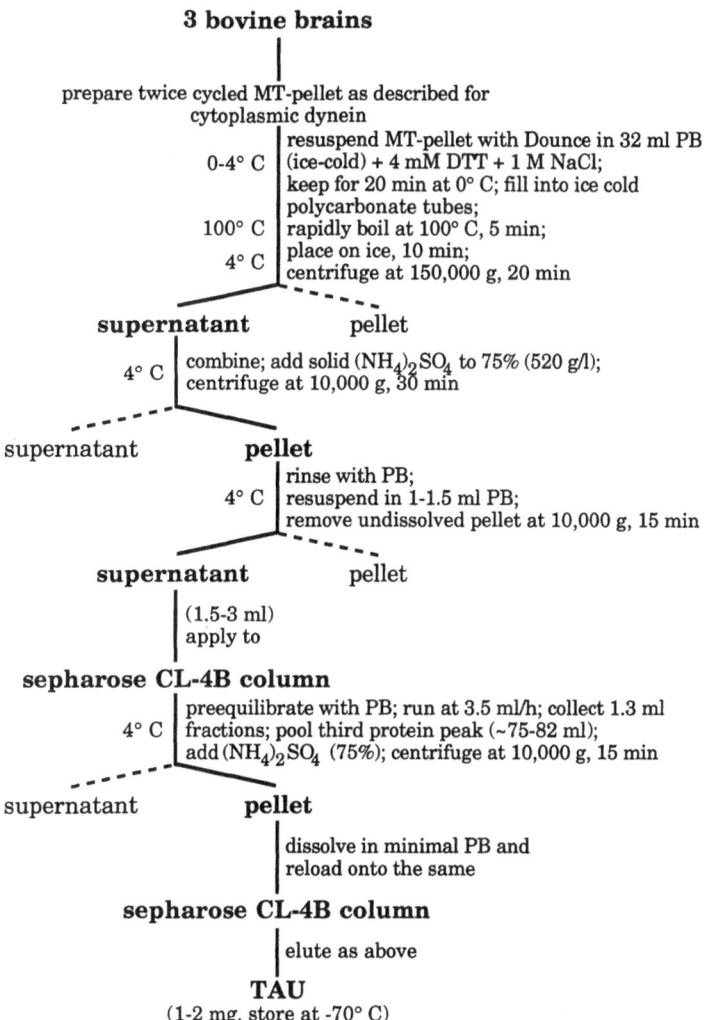

3 bovine brains

prepare twice cycled MT-pellet as described for
cytoplasmic dynein

0-4° C — resuspend MT-pellet with Dounce in 32 ml PB (ice-cold) + 4 mM DTT + 1 M NaCl; keep for 20 min at 0° C; fill into ice cold polycarbonate tubes;

100° C — rapidly boil at 100° C, 5 min;

4° C — place on ice, 10 min; centrifuge at 150,000 g, 20 min

supernatant   pellet

4° C — combine; add solid $(NH_4)_2SO_4$ to 75% (520 g/l); centrifuge at 10,000 g, 30 min

supernatant   **pellet**

4° C — rinse with PB; resuspend in 1-1.5 ml PB; remove undissolved pellet at 10,000 g, 15 min

supernatant   pellet

(1.5-3 ml)
apply to

sepharose CL-4B column

4° C — preequilibrate with PB; run at 3.5 ml/h; collect 1.3 ml fractions; pool third protein peak (~75-82 ml); add $(NH_4)_2SO_4$ (75%); centrifuge at 10,000 g, 15 min

supernatant   **pellet**

dissolve in minimal PB and
reload onto the same

sepharose CL-4B column

elute as above

TAU
(1-2 mg, store at -70° C)

TEKTINS

Tektins A (55 kDa), B (51 kDa) and C (47 kDa) are a protein family forming 2 nm fibrils, antigenically related to IF-proteins, associated with ciliary and flagellar MTs.

Source:	sea urchin
Equipment:	• sonifier (Branson)
	• centrifuge TL-100 (Beckman Instr.)
	• μ Bondapak C_{18} reverse-phase (Millipore Corp.) column connected to Waters system.
Chemicals:	Tris-HCl, KCl, $MgSO_4$, ATP, DTT, EDTA, Triton X-100, lysine, urea, sarkosyl, guanidinium-HCl, trifluoroacetic acid (TFA), acetonitrile
Have ready:	**extraction solution I:** 1% Triton X-100, 0.15 M KCl, 5 mM $MgSO_4$, 0.5 mM EDTA, 1 mM ATP, 1 mM DTT, 10 mM Tris-HCl, pH 8.3
	TED-buffer: 1 mM Tris, 0.1 mM EDTA, 0.5 mM DTT, pH 8.3
	extraction solution II: 50 mM Tris, 50 mM lysine, 1 mM EDTA, pH 8.3 + 0.5% sarkosyl, 2.0 M urea
	HPLC lysis buffer: 6 M guanidinium-HCl, 5 mM DTT, 10 mM Tris, pH 8.3
Reference:	Linck, R.W. and Stephens, R.E. (1987) J. Cell Biol. 104, 1069-1075

sea urchin sperm (10 mg protein /ml)

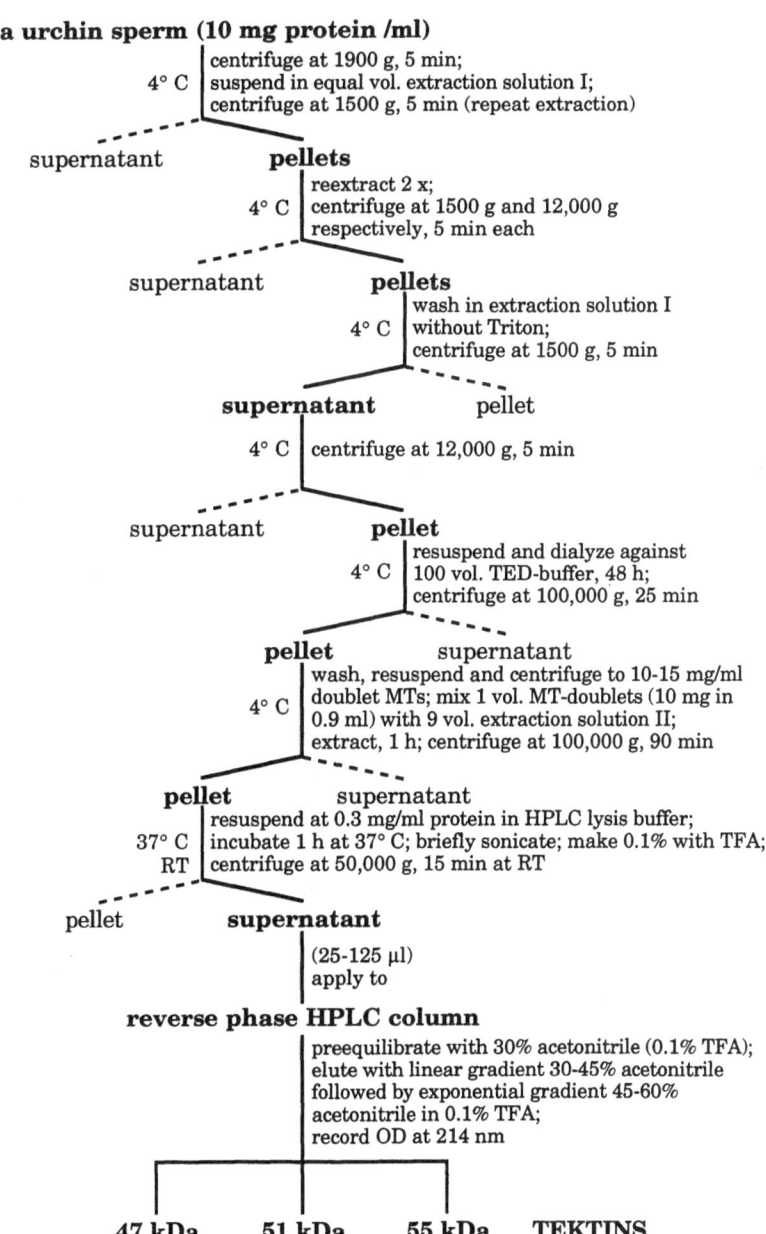

4° C | centrifuge at 1900 g, 5 min;
suspend in equal vol. extraction solution I;
centrifuge at 1500 g, 5 min (repeat extraction)

supernatant   **pellets**

4° C | reextract 2 x;
centrifuge at 1500 g and 12,000 g
respectively, 5 min each

supernatant   **pellets**

4° C | wash in extraction solution I
without Triton;
centrifuge at 1500 g, 5 min

supernatant   pellet

4° C | centrifuge at 12,000 g, 5 min

supernatant   **pellet**

4° C | resuspend and dialyze against
100 vol. TED-buffer, 48 h;
centrifuge at 100,000 g, 25 min

pellet   supernatant

4° C | wash, resuspend and centrifuge to 10-15 mg/ml
doublet MTs; mix 1 vol. MT-doublets (10 mg in
0.9 ml) with 9 vol. extraction solution II;
extract, 1 h; centrifuge at 100,000 g, 90 min

pellet   supernatant

37° C | resuspend at 0.3 mg/ml protein in HPLC lysis buffer;
RT | incubate 1 h at 37° C; briefly sonicate; make 0.1% with TFA;
centrifuge at 50,000 g, 15 min at RT

pellet   **supernatant**

(25-125 µl)
apply to

reverse phase HPLC column

preequilibrate with 30% acetonitrile (0.1% TFA);
elute with linear gradient 30-45% acetonitrile
followed by exponential gradient 45-60%
acetonitrile in 0.1% TFA;
record OD at 214 nm

47 kDa   51 kDa   55 kDa   TEKTINS

α/β TUBULIN

A 50 kDa heterodimer consisting of ~450 amino acids, constituating microtubules. One of the most important and widespread cytoskeletal proteins.

Source: calf brain

Equipment:
- centrifuge, GSA-rotor, SS-34 rotor
- polypropylene tubes
- Waring blender
- Dounce (teflon/glass) homogenizer
- water bath
- DEAE-sephadex A-50 (Pharmacia) column (4 ml)

Chemicals: PIPES-NaOH, EGTA, $MgSO_4$, NaCl, GTP, ATP, sucrose

Have ready: **PEM-buffer:** (1.1 l) 0.1 M PIPES-NaOH, pH 6.6 + 1 mM EGTA, 1 mM $MgSO_4$

sucrose underlayer: 90 ml PEM-buffer, 10% sucrose, 1 mM GTP

DEAE-column buffer: 100 ml PEM-buffer, 0.25 M NaCl, 0.1 mM GTP

Reference: Vallee, R.B. (1986) Methods Enzymol. 134, 89-104
Gen-Bank: more than 80 tubulin sequences in SWISS PROT-data bank

calf brain (5-6)

0-4° C | dissect and obtain 600 g cerebral cortex; homogenize in 900 ml PEM-buffer + 63 μl 2-mercaptoethanol with Waring blender at low speed;

2° C | homogenize further in 10-12 aliquots with 200-250 ml teflon/glass Dounce homogenizer, 2 passes at 200 rpm; centrifuge (6 x 290 ml) at 12,000 rpm in GSA-rotor, 90 min

supernatant   pellet

37° C | (800-850 ml) add GTP to 0.1 mM and ATP to 2.5 mM; incubate at 37° C, 25-30 min; layer on top of 20 ml sucrose underlayer; centrifuge at 12,000 rpm, 45 min

supernatant   **pellet**

0° C | resuspend in 75 ml ice-cold PEM-buffer + 1 mM GTP; homogenize in teflon/glass homogenizer

2° C | (2000 rpm, 2 passes); incubate on ice, 30 min; centrifuge at 18,000 rpm in SS-34 rotor, 30 min

supernatant   pellet

37° C | incubate in 8 x 12 ml centrifuge tubes, 10-15 min at 37° C; centrifuge at 18,000 rpm, 30 min;

supernatant   **pellet**

0° C | resuspend in 20 ml ice-cold PEM-buffer + 1 mM GTP; homogenize as above; incubate on ice; centrifuge at 18,000 rpm, 30 min

supernatant   pellet

37° C | warm up to 37° C, 10-15 min; centrifuge in 2 preweighed 12 ml polypropylene tubes at 18,000 rpm, 30 min;

supernatant   **pellet**

add 0.25 M NaCl in PEM-buffer; vortex; apply to

DEAE-sephadex column

4° C | preequilibrate with DEAE-column buffer; wash with 20 ml column buffer; elute with 10 ml PEM-buffer + 0.5 M NaCl, 0.1 mM GTP, at 0.5 ml/min; pool tubulin (~4 ml); dialyze against PEM-buffer + 0.1 mM GTP; store at -80° C in liquid nitrogen (500-1000 μl aliquots)

α/β TUBULIN
(7-10 mg/ml)

X-MAP

A 215 kDa MAP, an elongated monomer with 5 S sedimentation coefficient present in Xenopus eggs.

Source:	Xenopus laevis
Equipment:	• centrifuge, SW-50.1 rotor, Ti-50 rotor • phosphocellulose column (35 ml) • Mono Q HR 5/5 FPLC (Pharmacia) column • ultra filtration device • Biosil TSK 400 HPLC gel filtration (Bio-Rad) column (7.5 x 300 mm)
Chemicals:	K-PIPES, $MgCl_2$, EGTA, DTT, NaF, phospho-cellulose P-11 (Whatman), imidazole-HCl, NaCl, $(NH_4)_2SO_4$ **protease inhibitors:** PMSF, benzamidine-HCl, phenanthroline, pepstatin A
Have ready: protease benzamidine-HCl, phenanthroline	**BRB-buffer:** 80 mM K-PIPES, 1 mM $MgCl_2$, 1 mM EGTA, pH 7.0 + 1 mM DTT, 5 mM NaF + 1 x inhibitors: 0.1 mM PMSF, 0.1 mM 1 µg/ml pepstatin A, 1 µg/ml
	cold saturated $(NH_4)_2SO_4$
	MQ-buffer: 50 mM imidazole-HCl, 1 mM $MgCl_2$, 1 mM EGTA, 40 mM NaCl, pH 6.6 + 1 mM DTT + 1 x protease inhibitors
Reference:	Gard, D.L. and Kirschner, M.W. (1987) J. Cell Biol. 105, 2203-2215

Xenopus eggs (500-800 ml)

4° C — wash activated eggs 3-5 x in cold BRB-buffer; lyse after dejelling by pipetting through 25 ml pipette in 0.2 vol. buffer; centrifuge at 235,000 g, 90 min in SW 50.1 rotor; collect aqueous layer between yolk and lipid by puncturing

supernatant   pellet

4° C — (700-1000 mg, 10-15 mg/ml) batch-adsorb to P-11 (phosphocellulose) preequilibrated with BRB-buffer; incubate 3-12 h (30 mg protein/ml P-11); pour

phosphocellulose column

4° C — wash with BRB-buffer; step elute with 0.2, 0.3, 0.5 M NaCl in BRB-buffer; collect 200-300 mM eluate; add equal volume of cold $(NH_4)_2SO_4$ to 50%; centrifuge at 10,000 g, 15 min

supernatant   **pellet**

4° C — resuspend in MQ-buffer and dialyze, 3 h; clarify at 50,000 rpm in Ti-50 rotor, 45 min; apply to

Mono Q column

4° C — preequilibrated with MQ-buffer (0.3 ml/min); elute with linear 0-0.6 M NaCl in MQ-buffer at 0.5 ml/min; collect 0.5 ml fractions into 0.1 ml BRB + 5 x protease inhibitors; collect peak at 0.13 M NaCl; add 0.1 vol. of 2 M NaCl; concentrate to 150-200 µl by ultrafiltration (Centricon 10); apply to

Biosil TSK 400 HPLC column

4° C — preequilibrate with BRB + 200 mM NaCl; elute with 0.5 ml/min; collect 0.25 ml fractions into 50 µl BRB + 5 x protease inhibitors; pool protein eluting between 670-1000 kDa

X-MAP

(50-75 µg/ml, store at -70° C after adding 100µg/ml BSA)

III Motor Proteins

AXONEMAL DYNEIN

A 20 S two headed microtubular based motor ATP-ase of ~1,300 kDa consisting of heavy (>400 kDa), intermediate (75-120 kDa) and light chains (15-30 kDa). Located in cilia and flagella and responsible for bending motions.

Source:	sea urchin
Equipment:	• Dounce homogenizer • centrifuge • SW-41 rotor • sucrose gradient • SDS-PAGE (3-6% gels)
Chemicals:	sucrose, imidazole-HCl, NaCl, MgSO$_4$, CaCl$_2$, EDTA, 2-mercaptoethanol, DTT
Have ready:	**sea water** **isolation buffer:** 0.1 M NaCl, 5 mM imidazole-HCl, 4 mM MgSO$_4$, 1 mM CaCl$_2$, 1 mM EDTA, 7 mM 2-mercaptoethanol, pH 7.0 **high salt buffer:** 0.6 M NaCl, 5 mM imidazole-HCl, 4 mM MgSO$_4$, 1 mM CaCl$_2$, 1 mM EDTA, 7 mM 2-mercaptoethanol, 1 mM DTT, pH 7.0
Reference:	Bell, C.W. et al. (1982) Methods Enzymol. 85, 450-467 Gen-Bank: D01021, X59603 (sea urchin)

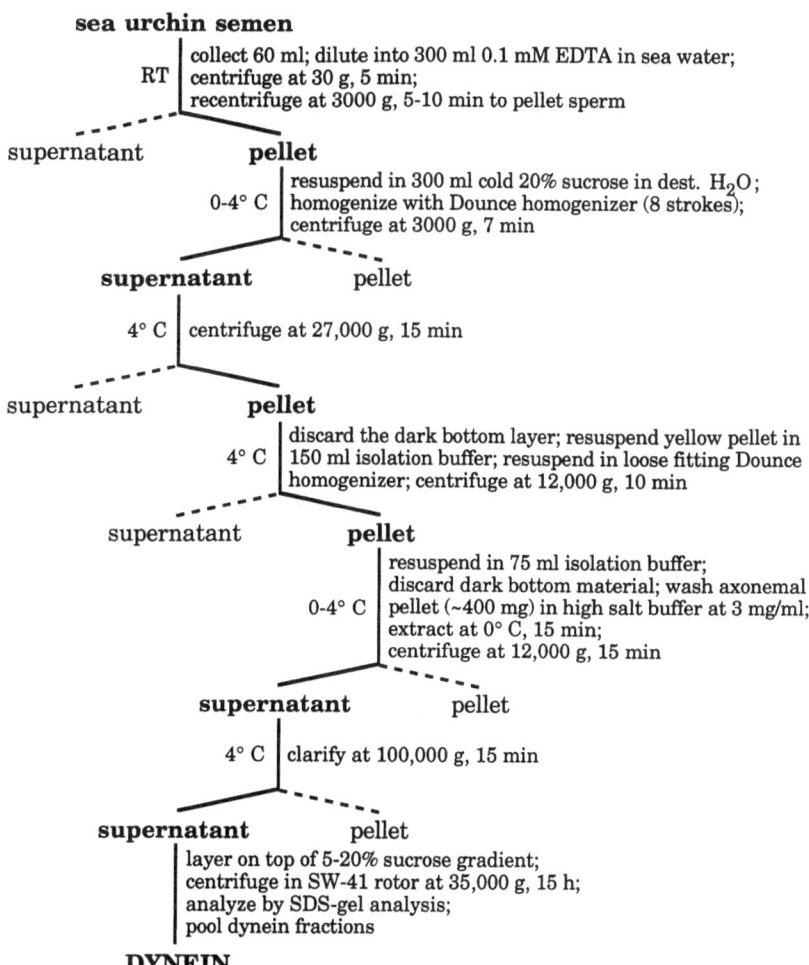

sea urchin semen

RT | collect 60 ml; dilute into 300 ml 0.1 mM EDTA in sea water;
centrifuge at 30 g, 5 min;
recentrifuge at 3000 g, 5-10 min to pellet sperm

supernatant   **pellet**

0-4° C | resuspend in 300 ml cold 20% sucrose in dest. H_2O;
homogenize with Dounce homogenizer (8 strokes);
centrifuge at 3000 g, 7 min

supernatant   pellet

4° C | centrifuge at 27,000 g, 15 min

supernatant   **pellet**

4° C | discard the dark bottom layer; resuspend yellow pellet in
150 ml isolation buffer; resuspend in loose fitting Dounce
homogenizer; centrifuge at 12,000 g, 10 min

supernatant   **pellet**

0-4° C | resuspend in 75 ml isolation buffer;
discard dark bottom material; wash axonemal
pellet (~400 mg) in high salt buffer at 3 mg/ml;
extract at 0° C, 15 min;
centrifuge at 12,000 g, 15 min

supernatant   pellet

4° C | clarify at 100,000 g, 15 min

supernatant   pellet

layer on top of 5-20% sucrose gradient;
centrifuge in SW-41 rotor at 35,000 g, 15 h;
analyze by SDS-gel analysis;
pool dynein fractions

DYNEIN

CALTRACTIN

A 20 kDa Ca^{++}-binding phosphoprotein from the basal body complex of Chlamydomonas involved in disposition and segregation of the basal body and its homologue - the microtubule organizing center (MTOC).

Source: Tetraselmis

Equipment:
- centrifuge
- sonicator
- Dounce homogenizer (tight)
- phenyl-sepharose column (Pharmacia) (10 x 1 cm)

Chemicals: Tris-HCl, EGTA, $CaCl_2$, KCl, ammonium-bicarbonate
protease inhibitors: PMSF

Have ready: **lysis buffer:** 20 mM Tris-HCl, 1 mM EGTA, 5 mM PMSF, pH 7.4

PS-column buffer: lysis buffer + 0.5 M KCl, 2 mM $CaCl_2$, pH 7.4

Reference: Salisbury et al. (1986) Methods Enzymol. 134, 408-414
Gen-Bank: 12634

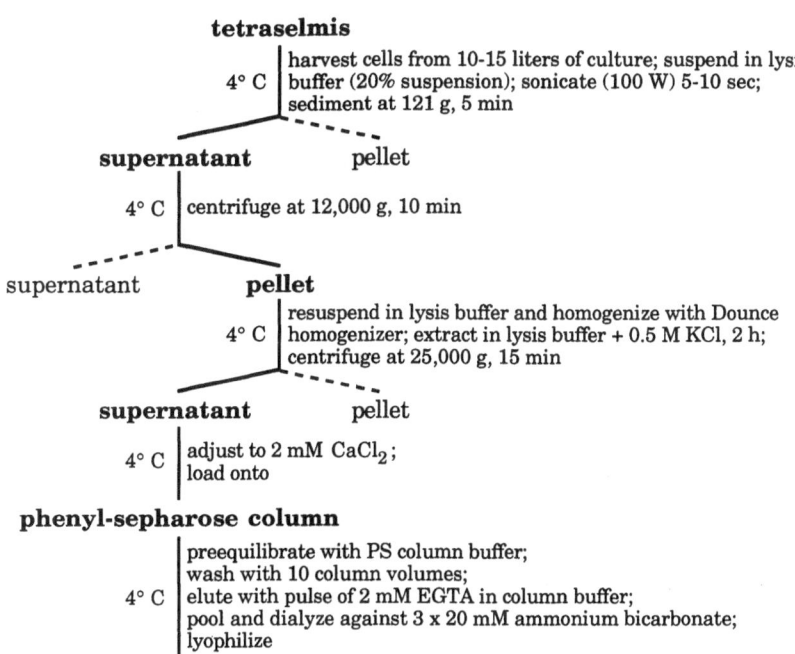

tetraselmis

4° C | harvest cells from 10-15 liters of culture; suspend in lysis buffer (20% suspension); sonicate (100 W) 5-10 sec; sediment at 121 g, 5 min

supernatant   pellet

4° C | centrifuge at 12,000 g, 10 min

supernatant   **pellet**

4° C | resuspend in lysis buffer and homogenize with Dounce homogenizer; extract in lysis buffer + 0.5 M KCl, 2 h; centrifuge at 25,000 g, 15 min

supernatant   pellet

4° C | adjust to 2 mM $CaCl_2$; load onto

phenyl-sepharose column

4° C | preequilibrate with PS column buffer; wash with 10 column volumes; elute with pulse of 2 mM EGTA in column buffer; pool and dialyze against 3 x 20 mM ammonium bicarbonate; lyophilize

CALTRACTIN

CYTOPLASMIC MYOSIN II

A two-headed ATP hydrolyzing motor protein consisting of two
heavy chains (up to 240 kDa) and two pairs of light chains (15-20 kDa
each), 200 nm in length forming filaments and soluble monomers.

Source: platelets

Equipment:
- centrifuge
- 4% argarose gel filtration column (2.6 x 90 cm)
  Bio Rad A-15m (200-400 mesh)
- Dounce homogenizer

Chemicals: KCl, MgCl$_2$, Na$_4$P$_2$O$_7$, imidazole, DTT, acetic acid,
                 ATP, KI, (NH$_4$)$_2$SO$_4$

Have ready: **extraction solution:** 0.9 M KCl, 15 mM Na-
                 pyrophosphate, 30 mM imidazole, 5 mM MgCl$_2$, 3 mM
                 DTT, pH 7.0

                 ice-cold 2 mM MgCl$_2$

                 0.5 M acetic acid

                 KI-ATP buffer: 0.6 M KI, 5 mM ATP, 5 mM DTT,
                 1 mM MgCl$_2$, 20 mM imidazole, pH 7.0

                 cold saturated (NH$_4$)$_2$SO$_4$

                 gel-filtration buffer: (600 ml) 0.6 M KCl, 1 mM
                 DTT, 10 mM imidazole, pH 7.0

Reference: Pollard, T.D. (1982) Methods Enzymol. 85, 331-356
                 Gen-Bank: M14628 (Dictyostelium), Y00624, M12702,
                 M12703, M11938, M11938 (Acanthamoeba), M31013
                 (human), M26510 (chicken), X53947 (yeast)

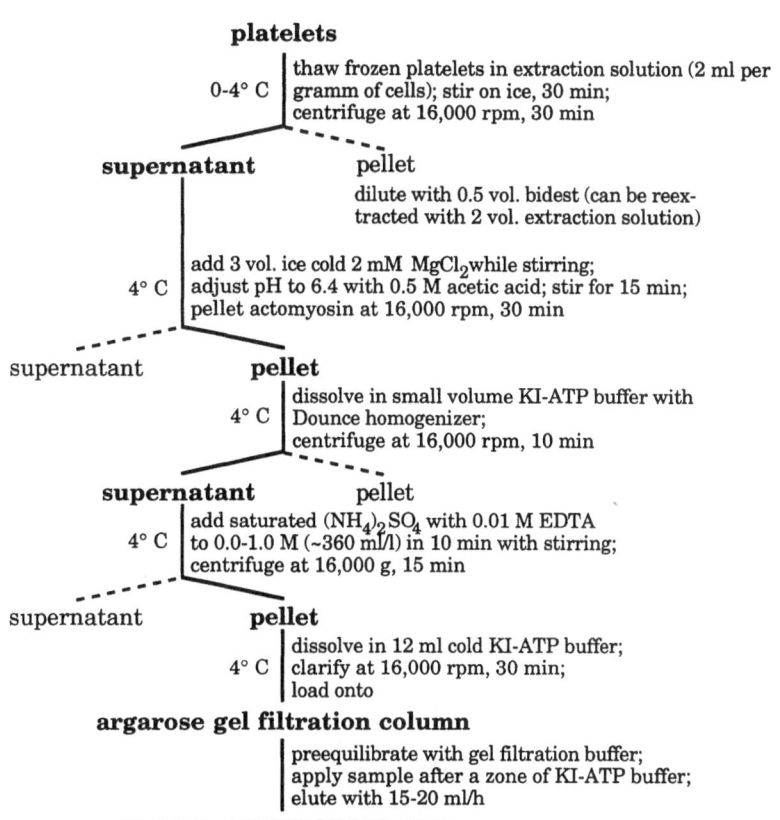

platelets

0-4° C | thaw frozen platelets in extraction solution (2 ml per gramm of cells); stir on ice, 30 min; centrifuge at 16,000 rpm, 30 min

supernatant   pellet

dilute with 0.5 vol. bidest (can be reextracted with 2 vol. extraction solution)

4° C | add 3 vol. ice cold 2 mM $MgCl_2$ while stirring; adjust pH to 6.4 with 0.5 M acetic acid; stir for 15 min; pellet actomyosin at 16,000 rpm, 30 min

supernatant   **pellet**

4° C | dissolve in small volume KI-ATP buffer with Dounce homogenizer; centrifuge at 16,000 rpm, 10 min

supernatant   pellet

4° C | add saturated $(NH_4)_2SO_4$ with 0.01 M EDTA to 0.0-1.0 M (~360 ml/l) in 10 min with stirring; centrifuge at 16,000 g, 15 min

supernatant   **pellet**

4° C | dissolve in 12 ml cold KI-ATP buffer; clarify at 16,000 rpm, 30 min; load onto

argarose gel filtration column

preequilibrate with gel filtration buffer; apply sample after a zone of KI-ATP buffer; elute with 15-20 ml/h

CYTOPLASMIC MYOSIN II
(store on ice with 0.02% NaN_3 for 5-10 days)

189

DYNACTIN

A 150 kDa activator protein of dynein with two highly conserved α-helical coiled-coil domains. A mediator for dynein-dependent retrograde transport.

Source: chicken brain

Equipment:
- centrifuge, SA-600, SW-50.1, SW-40 rotors
- 0.2 μm filter
- Dounce homogenizer
- water bath
- sucrose gradients
- Mono Q 4R5/5 column
- Amicon 30 concentrator

Chemicals: K-PIPES, $MgSO_4$, EDTA, EGTA, DTT, ATP, GTP, taxol, AMP-PNP, sucrose, KCl
protease inhibitors: 1 mM PMSF, 1 μg/ml pepstatin A, 10 μg/ml TAME, 10 μg/ml BAME (N-α-benzoyl-L-arginine methyl ester), 1 μg/ml leupeptin, 10 μg/ml soybean trypsin inhibitor

Have ready: **homogenization buffer:** 35 mM K-PIPES, 5 mM $MgSO_4$, 1 mM EGTA, 0.5 mM EDTA, 1 mM DTT, pH 7.2 + protease inhibitors

MT-polymerization solution: 0.5 mM ATP, 1 mM GTP, 20 μM taxol, (final conc.)

polymerized microtubules: tubulin (0.25 mg/ml) + 1 mM GTP + 20 μM taxol, polymerized for 15 min at 37° C

sucrose step gradient: 12.5 and 25% sucrose in homogenization buffer + 1 mM GTP + 20 μM taxol

ATP-release buffer: homogenization buffer + 10 mM ATP + 5 mM (additional) $MgSO_4$, 1 mM GTP, 20 μM taxol, pH 7.2

linear sucrose gradient: 5-20% in homogenization buffer + 0.5 mM ATP, 1 mM DTT (use catalase 11.3S and thyroglobulin 19S as standards)

Reference: Schroer, T.A. and Sheetz, M.P. (1991) J. Cell Biol. 115, 1309-1318
Gen-Bank: X62773 (chicken)

chicken brain (~20 g)

4° C | homogenize with Dounce homogenizer in 20 ml homogenization buffer (10 strokes); let stand, 5 min; rehomogenize (10 strokes); centrifuge at 16,500 rpm in SA-600 rotor, 30 min

supernatant   pellet

4° C | centrifuge at 182,000 g, 30 min

supernatant   pellet

37° C | add MT-polymerization solution; incubate 30 min;
25° C | centrifuge at 182,000 g, 30 min

supernatant   pellet

37° C | make 4 mM AMP-PNP and MgSO$_4$; add microtubules polymerized from 0.25 mg/ml tubulin; incubate, 20 min; layer on top of two step 12.5 and 25% sucrose gradient (5-10 ml);
25° C | centrifuge in 39 ml SW 28 tube at 105,000 g, 45 min

supernatant   **pellet**

RT | gently resuspend by slowly adding ATP-release buffer (0.1 x original supernatant volume); incubate at RT, 15 min;
25° C | centrifuge in SW 50.1 tubes at 188,000 g, 30 min

supernatant   pellet

4° C | (ATP release: 2.5-3 ml total) layer on top of 11.5 ml 5-20% sucrose gradient; centrifuge in SW 40 rotor at 145,000 g, 13.5 h; pool 20 S fractions; pass through 0.2 µm filter; load onto

HR5/5 Mono Q column

| wash with homogenization buffer; elute with 15 ml 0-250 mM linear KCl gradient and 25 ml 250-400 mM linear KCl gradient; collect 1 ml fractions at 1 ml/min; desalt and concentrate with Amicon 30 concentrator

DYNACTIN
(50-150 µl, ~200 µg/ml)

191

DYNAMIN

A 100 kDa microtubule associated nucleotide sensitive motor protein hydrolyzing GTP.

Source: calf brain

Equipment:
- teflon homogenizer
- water bath
- centrifuge
- GSA-rotor, Ti-60 rotor, SS-34 rotor
- 4 PD-10 gel filtration columns (Pharmacia)
- DEAE-sepharose CL-6B column (0.3 ml)
- prepacked NAP-10 gel filtration column (Pharmacia)

Chemicals: PIPES-NaOH, HEPES, $MgCl_2$, EDTA, DTT, sucrose, taxol, AMP-PNP, Mg-GTP, $MgSO_4$, Na/PO_4, sodium glutamate, DMSO
protease inhibitors: PMSF, TAME, leupeptin, pepstatin A

Have ready: **extraction solution:** 250 ml 0.05 M PIPES-NaOH, 0.05 M HEPES, 2 mM $MgCl_2$, 1 mM EDTA, 1 mM PMSF, 10 µg/ml leupeptin, 10 µg/ml tosyl-arginine methyl ester (TAME), 1 µg/ml pepstatin A, 1 mM DTT, pH 7.0

taxol, 350 µl of 10 mM stock in DMSO, stored at -80° C

sucrose cushion: 40 ml extraction solution + 7.5% sucrose

Mg-AMP-PNP: 0.3 ml of 0.1 M stock made equimolar with $MgSO_4$, pH 7.0 (with NaOH)

Mg-GTP: 0.3 ml of 0.1 M stock made equimolar with $MgSO_4$, pH 7.0 (with NaOH)

P/G-buffer: (250 ml) 5 mM sodium phosphate, 50 mM sodium glutamate, 1 mM $MgSO_4$, 1 mM DTT, pH 7.0

DEAE-purified tubulin (~5 mg/ml), 0.3 ml in PEM-buffer: 100 mM PIPES, 1 mM $MgSO_4$, 1 mM EGTA, 0.1 mM GTP, pH 6.6

prepare mictotubules by adding 3 µl taxol (10 µM final conc.) to 300 µl tubulin, incubate at 37° C, 5 min, centrifuge at 18,000 rpm (SS-34 rotor), 30 min, at 37° C

Reference: Shpenter, H.S. and Vallee, R.B. (1991) Methods Enzymol. 196, 192-201
Gen-Bank: X54531 (rat brain), X59448, X59449 (Drosophila)

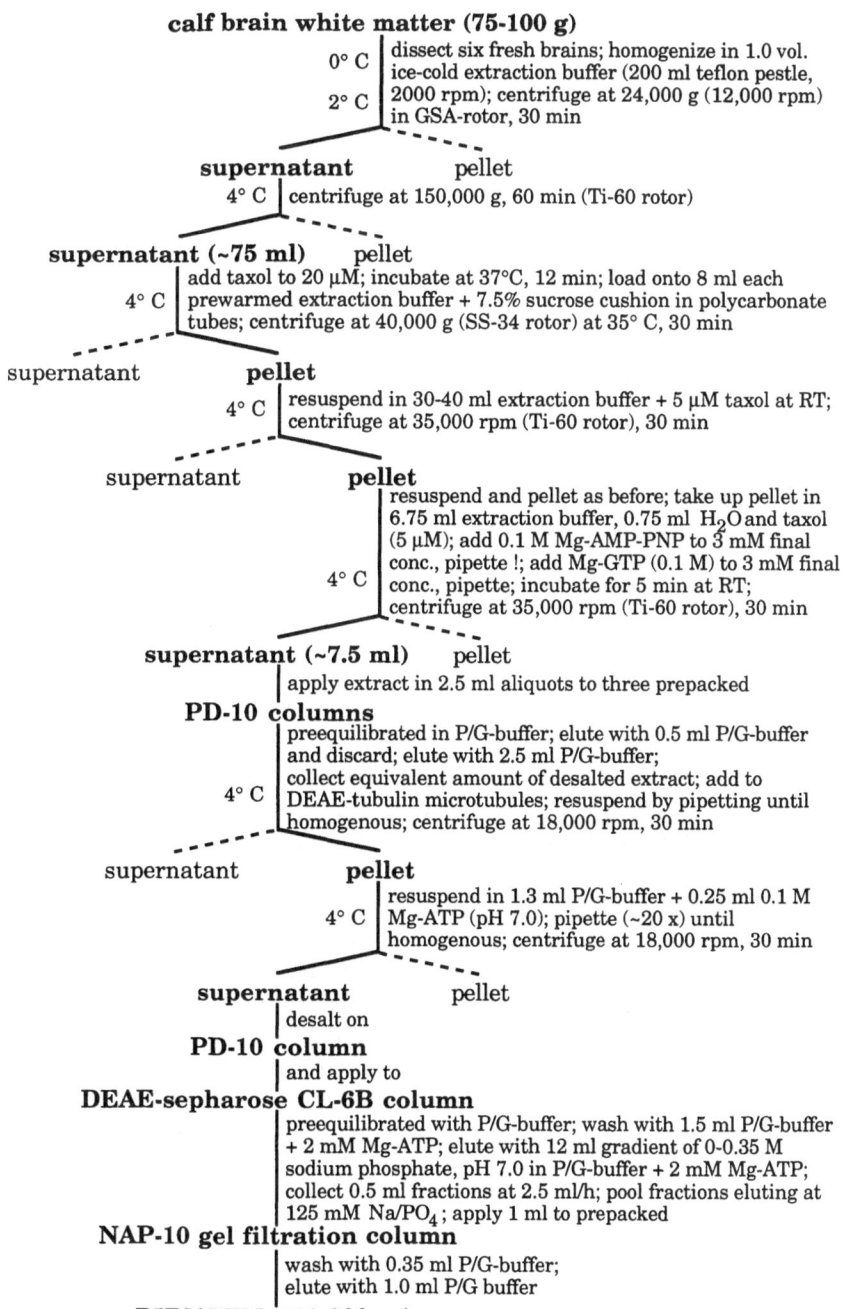

calf brain white matter (75-100 g)

0° C | dissect six fresh brains; homogenize in 1.0 vol.
ice-cold extraction buffer (200 ml teflon pestle,
2° C | 2000 rpm); centrifuge at 24,000 g (12,000 rpm)
in GSA-rotor, 30 min

supernatant   pellet

4° C | centrifuge at 150,000 g, 60 min (Ti-60 rotor)

supernatant (~75 ml)   pellet

4° C | add taxol to 20 µM; incubate at 37°C, 12 min; load onto 8 ml each
prewarmed extraction buffer + 7.5% sucrose cushion in polycarbonate
tubes; centrifuge at 40,000 g (SS-34 rotor) at 35° C, 30 min

supernatant   **pellet**

4° C | resuspend in 30-40 ml extraction buffer + 5 µM taxol at RT;
centrifuge at 35,000 rpm (Ti-60 rotor), 30 min

supernatant   **pellet**

4° C | resuspend and pellet as before; take up pellet in
6.75 ml extraction buffer, 0.75 ml H$_2$O and taxol
(5 µM); add 0.1 M Mg-AMP-PNP to 3 mM final
conc., pipette !; add Mg-GTP (0.1 M) to 3 mM final
conc., pipette; incubate for 5 min at RT;
centrifuge at 35,000 rpm (Ti-60 rotor), 30 min

supernatant (~7.5 ml)   pellet

| apply extract in 2.5 ml aliquots to three prepacked

PD-10 columns

4° C | preequilibrated in P/G-buffer; elute with 0.5 ml P/G-buffer
and discard; elute with 2.5 ml P/G-buffer;
collect equivalent amount of desalted extract; add to
DEAE-tubulin microtubules; resuspend by pipetting until
homogenous; centrifuge at 18,000 rpm, 30 min

supernatant   **pellet**

4° C | resuspend in 1.3 ml P/G-buffer + 0.25 ml 0.1 M
Mg-ATP (pH 7.0); pipette (~20 x) until
homogenous; centrifuge at 18,000 rpm, 30 min

supernatant   pellet

| desalt on

PD-10 column

| and apply to

DEAE-sepharose CL-6B column

| preequilibrated with P/G-buffer; wash with 1.5 ml P/G-buffer
+ 2 mM Mg-ATP; elute with 12 ml gradient of 0-0.35 M
sodium phosphate, pH 7.0 in P/G-buffer + 2 mM Mg-ATP;
collect 0.5 ml fractions at 2.5 ml/h; pool fractions eluting at
125 mM Na/PO$_4$; apply 1 ml to prepacked

NAP-10 gel filtration column

| wash with 0.35 ml P/G-buffer;
elute with 1.0 ml P/G buffer

DYNAMIN (100-200 µg)

KINESIN

A tetramer consisting of two heavy chains of 120 kDa and two light chains of 62 kDa. A two headed elongated molecule of 80 nm in length, associated with microtubules, and acting as anterograde force generating enzyme.

Source: bovine brain

Equipment:
- centrifuge
- GSA-, Ti-45, Ti-60, SA-600, SW-41 rotors
- ultrafiltration device
- Waring blender
- water bath
- two ultra centrifuges, two SW-28 rotors
- Toyo pearl HW-65F gel filtration column (4.8 x 120 cm) (Supelco, Bellefonte, PA)
- S-sepharose column (Pharmacia) (2.5 x 3 cm)
- hydroxylapatide column (0.5-1 ml)
- sucrose gradient

Chemicals: PIPES, EGTA, $MgSO_4$, GTP, ATP,
2-mercaptoethanol, taxol, AMP-PNP, sucrose,
$(NH_4)_2SO_4$, $MgCl_2$, imidazole, glycerol, NaCl, Na/PO_4
protease inhibitors: PMSF, leupeptin, pepstatin A

Have ready: **extraction solution:** 0.1 M PIPES, 1 mM EGTA,
1 mM $MgSO_4$, 0.1 mM GTP, 0.1 mM ATP, 10 mM
2-mercaptoethanol, pH 6.62 + 1 µg/ml leupeptin,
1 µg/ml pepstatin A, 1 mM PMSF

pure tubulin (see tubulin purification)

sucrose cushion: 0.1 M PIPES, 1 mM EGTA, 1 mM
$MgSO_4$, 0.5 mM AMP-PNP, 20% sucrose, pH 6.62 +
protease inhibitors 1 µg/ml each

ATP-buffer: 0.1 M PIPES, 1 mM EGTA, 1 mM
$MgSO_4$, 10 mM 2-mercaptoethanol, 5 µM taxol, 5 mM
ATP, pH 6.62 + protease inhibitors 1 µg/ml each

ice-cold saturated $(NH_4)_2SO_4$

IMEG-buffer: 15 mM imidazole, 1 mM EGTA, 2 mM
$MgCl_2$, 10% glycerol, pH 7.0

IME-buffer: IMEG-buffer without glycerol

sodium phosphate buffer: 20 mM sodium
phosphate, 0.15 M NaCl, pH 6.8

Reference: Wagner, M.C. et al. (1991) Methods Enzymol. 196, 157-179
Gen-Bank: kinesin heavy chains: M24441
(drosophila), J05258 (squid), X65873 (human)
kinesin light chains: M75146, M75147, M75148 (rat)

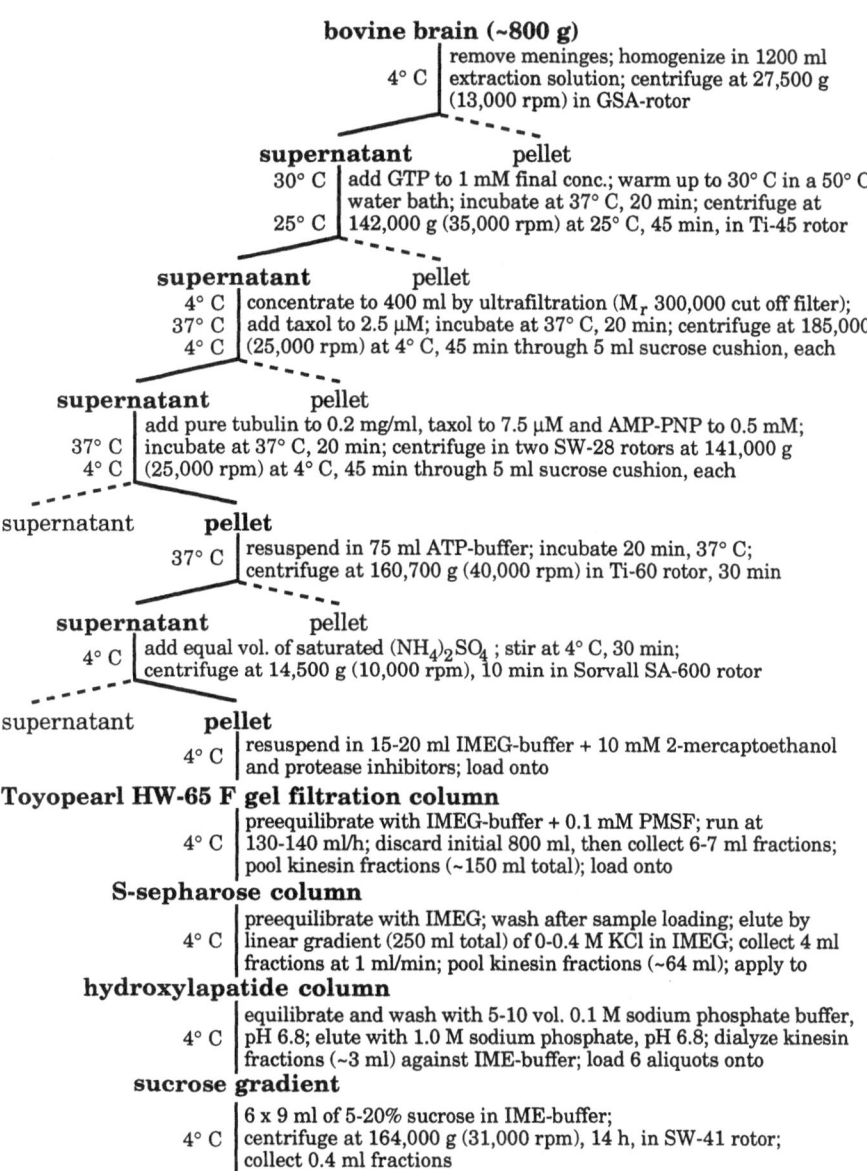

bovine brain (~800 g)

4° C | remove meninges; homogenize in 1200 ml extraction solution; centrifuge at 27,500 g (13,000 rpm) in GSA-rotor

supernatant   pellet

30° C | add GTP to 1 mM final conc.; warm up to 30° C in a 50° C water bath; incubate at 37° C, 20 min; centrifuge at
25° C | 142,000 g (35,000 rpm) at 25° C, 45 min, in Ti-45 rotor

supernatant   pellet

4° C | concentrate to 400 ml by ultrafiltration (M_r 300,000 cut off filter);
37° C | add taxol to 2.5 μM; incubate at 37° C, 20 min; centrifuge at 185,000 g
4° C | (25,000 rpm) at 4° C, 45 min through 5 ml sucrose cushion, each

supernatant   pellet

37° C | add pure tubulin to 0.2 mg/ml, taxol to 7.5 μM and AMP-PNP to 0.5 mM; incubate at 37° C, 20 min; centrifuge in two SW-28 rotors at 141,000 g
4° C | (25,000 rpm) at 4° C, 45 min through 5 ml sucrose cushion, each

supernatant   **pellet**

37° C | resuspend in 75 ml ATP-buffer; incubate 20 min, 37° C; centrifuge at 160,700 g (40,000 rpm) in Ti-60 rotor, 30 min

supernatant   pellet

4° C | add equal vol. of saturated $(NH_4)_2SO_4$; stir at 4° C, 30 min; centrifuge at 14,500 g (10,000 rpm), 10 min in Sorvall SA-600 rotor

supernatant   **pellet**

4° C | resuspend in 15-20 ml IMEG-buffer + 10 mM 2-mercaptoethanol and protease inhibitors; load onto

Toyopearl HW-65 F gel filtration column

4° C | preequilibrate with IMEG-buffer + 0.1 mM PMSF; run at 130-140 ml/h; discard initial 800 ml, then collect 6-7 ml fractions; pool kinesin fractions (~150 ml total); load onto

S-sepharose column

4° C | preequilibrate with IMEG; wash after sample loading; elute by linear gradient (250 ml total) of 0-0.4 M KCl in IMEG; collect 4 ml fractions at 1 ml/min; pool kinesin fractions (~64 ml); apply to

hydroxylapatide column

4° C | equilibrate and wash with 5-10 vol. 0.1 M sodium phosphate buffer, pH 6.8; elute with 1.0 M sodium phosphate, pH 6.8; dialyze kinesin fractions (~3 ml) against IME-buffer; load 6 aliquots onto

sucrose gradient

4° C | 6 x 9 ml of 5-20% sucrose in IME-buffer; centrifuge at 164,000 g (31,000 rpm), 14 h, in SW-41 rotor; collect 0.4 ml fractions

KINESIN (~2 mg)

MYOSIN I

Myosin I, a 110 kDa-calmodulin complex which binds to membranes, exhibits actin activated ATP-ase and produces movement.

Source: intestines of four white leghorn chickens

Equipment:
- centrifuge
- Waring blender
- glass-teflon-homogenizer with pestle
- sepharose CL-4B column (5 x 90 cm)
- S-sepharose column (2.5 x 10 cm)
- Mono Q FPLC column (0.5 x 5 cm)
- electrophoretic gels
- chemical hood

Chemicals: KH_2PO_4, NaCl, DTT, NaN_3, Na/PO$_4$, EDTA, EGTA, imidazole, KCl, $MgCl_2$, ATP, sucrose, TES
protease inhibitors: 0.2 mM PMSF, 1 mM DFP, 1 mg/ml leupeptin, 5 mg/ml pepstatin, 5 mg/ml aprotinin

Have ready: **buffer A:** 10 mM KH_2PO_4, 0.15 mM NaCl, 1 mM DTT, 0.02% (w/v) NaN_3, pH 7.5 + protease inhibitors

buffer B: 76 mM Na/PO$_4$, 19 mM KH_2PO_4, 12 mM EDTA, 1 mM DTT, 0.02% (w/v) NaN_3, pH 7.0 + protease inhibitors

buffer C: 76 mM Na/PO$_4$, 19 mM KH_2PO_4, 12 mM EDTA, 0.02% (w/v) NaN_3, pH 7.0

buffer D: 10 mM imidazole-HCl, 4 mM EDTA, 1 mM EGTA, 1 mM DTT, 0.02% (w/v) NaN_3, pH 7.3 + protease inhibitors

buffer E: 10 mM imidazole-HCl, 75 mM KCl, 5 mM $MgCl_2$, 1 mM EGTA, 1 mM DTT, 0.02% (w/v) NaN_3, pH 7.3 + protease inhibitors

buffer F: 10 mM imidazole-HCl, 0.2 M KCl, 5 mM $MgCl_2$, 5 mM ATP, 1 mM EGTA, 1 mM DTT, 0.02% NaN_3, pH 6.8 + protease inhibitors

buffer G: 10 mM TES, 0.3 M KCl, 1 mM EDTA, 5% (w/v) sucrose, 1 mM ATP, 5 mg/ml aprotinin A, 1 mg/ml leupeptin, 5 mg/ml pepstatin, 0.2 mM PMSF, 1 mM DTT, 0.02 % (w/v) NaN_3, pH 7.5

buffer H: 10 mM imidazole-HCl, 25 mM NaCl, 1 mM EDTA, 1 mM EGTA, 10% (w/v) sucrose, 0.2 mM PMSF, 1 mM DTT, pH 7.5

Reference: Swanljung-Collins, H. and Collins, J.H. (1991) Methods Enzymol. 196, 3-11

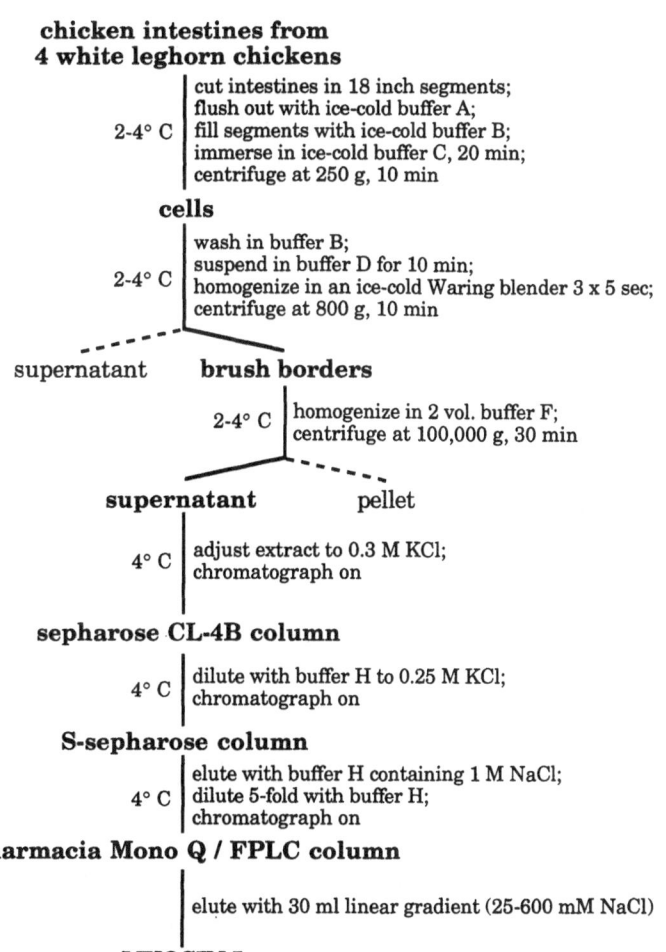

**chicken intestines from
4 white leghorn chickens**

2-4° C | cut intestines in 18 inch segments;
flush out with ice-cold buffer A;
fill segments with ice-cold buffer B;
immerse in ice-cold buffer C, 20 min;
centrifuge at 250 g, 10 min

cells

2-4° C | wash in buffer B;
suspend in buffer D for 10 min;
homogenize in an ice-cold Waring blender 3 x 5 sec;
centrifuge at 800 g, 10 min

supernatant   **brush borders**

2-4° C | homogenize in 2 vol. buffer F;
centrifuge at 100,000 g, 30 min

supernatant   pellet

4° C | adjust extract to 0.3 M KCl;
chromatograph on

sepharose CL-4B column

4° C | dilute with buffer H to 0.25 M KCl;
chromatograph on

S-sepharose column

4° C | elute with buffer H containing 1 M NaCl;
dilute 5-fold with buffer H;
chromatograph on

Pharmacia Mono Q / FPLC column

elute with 30 ml linear gradient (25-600 mM NaCl)

MYOSIN I
(~4.5 mg from 120 ml brush border extract)

PROTOZOAN MYOSIN I

A single headed mechanochemical enzyme with a native molecular mass of 140-160 kDa. A globular actin activated Mg^{++}-ATP-ase with a proposed key role for intracellular movement and amoeboid motility.

Source:	Acanthamoeba
Equipment:	• centrifuge, GSA-rotor
	• Dounce homogenizer
	• DE-52 ion echange column (10 x 25 cm)
	• phosphocellulose (P-11, Whatman) column (5 x 20 cm)
	• ATP-argarose (type 4, Pharmacia) column (1.6 x 12.5 cm)
	• Mono Q column (Pharmacia) (8 ml)
Chemicals:	imidazole-HCl, KCl, $Na_4P_2O_7$, DTT, Tris-HCl, EGTA, Trizma base, NaN_3, imidazole hydrochloride, EDTA
	protease inhibitors: PMSF (250 mM in ethanol), leupeptin, soybean trypsin inhibitor, pepstatin A, DFP, glycerol ultrapure
Have ready:	**extraction buffer (2 l/kg cells):** 30 mM imidazole hydrochloride, 75 mM KCl, 12 mM $Na_4P_2O_7$, pH 7.5 + 5 mM DTT, 0.5 mM PMSF, 2 mg/l leupeptin, 20 mg/l soybean trypsin inhibitor, 10 mg/l pepstatin A
	DE-dialysis buffer (30 l): 25 mM Tris-HCl, 7.5 mM $Na_4P_2O_7$, 0.5 mM DTT, 0.5 mM PMSF, pH 8.0
DTT,	**DE-buffer:** 25 mM Tris-HCl, 10 mM KCl, 1 mM pH 8.0 (make 4 l 10x concentrated stock)
	2 M imidazole hydrochloride, pH 7.5
concentrated	**PC-buffer:** 50 mM imidazole hydrochloride, 0.5 mM EGTA, 1 mM DTT, pH 7.5 (make 3 l 10x stock)
	ATP-Ag-buffer: 15 mM Tris-HCl, 50 mM KCl, 1 mM DTT, pH 7.5 (make 1 l 10x concentrated stock)
	Q-buffer: 25 mM Tris-HCl, 50 mM KCl, 1 mM DTT, pH 8.8 (make 1 l 10x concentrated stock)
	MI-storage buffer (1 l): 10 mM Tris-HCl, 100 mM KCl, 1 mM DTT, 0.01% NaN_3, 50% glycerol ultrapure, pH 7.5
Reference:	Lynch, T.J. et al. (1990) Methods Enzymol. 196, 12-23

axenic culture (900-1300 g packed cells)

4° C | homogenize with Dounce homogenizer (15 strokes, B pestle) in 2 vol. extraction buffer; centrifuge at 23,000 g (12,000 rpm), 1 h, in GSA-rotor

supernatant   pellet

4° C | remove lipid; add DFP to 0.5 mM; clarify at 100,000 g, 3.5 h (recentrifuge turbid lower supernatants, if necessary)

supernatant   pellet

4° C | adjust pH to 8.0 by adding 2 M Tris-base; dialyze overnight against DE-dialysis buffer; clarify at 23,000 g, 1 h

supernatant   pellet

load onto

DE-52 ion exchange column

4° C | equilibrate with DE-buffer; wash with 1-2 l 50 mM KCl in DE-buffer; elute myosin I with 1-2 l 125 mM KCl in DE-buffer; collect 20-25 ml fractions; pool and make 50 mM imidazole hydrochloride, pH 7.5, by addition of 2 M stock; add protease inhibitors; load onto

phosphocellulose column

4° C | equilibrated with PC-buffer; wash with PC-buffer; elute with 2.5 l linear 0-600 mM KCl gradient in PC-buffer; collect 20 ml fractions; proceed with myosin IB, myosin IC and myosin IA separately; pool and dialyze overnight against ATP-Ag-buffer + 20% glycerol; load onto

ATP-argarose column

4° C | preequilibrated with ATP-Ag-buffer; wash with 1-2 column volumes; elute with 400 ml linear 0-1 M KCl, 1 mM EDTA in ATP-Ag buffer; dialyze single peaks of myosin isozymes against 20 vol. Q-buffer + 100 mM KCl overnight; load onto

Mono Q column

4° C | preequilibrate with Q-buffer; elute with 50 ml linear gradient 160-275 mM KCl in Q-buffer; pool; dialyze against MI-storage buffer; keep at -20° C

MYOSIN I (A,B,C)
(2-5 mg)

199

SKELETAL MUSCLE MYOSIN

A two headed hexomeric actin-activated ATP-ase consisting of two heavy chains of ~200 kDa and four light chains of ~20 kDa. The tail portion consists of an α-helical coiled-coil rod of 150 nm in length. The most important mechano-enzyme for muscle contraction.

Source:	rabbit skeletal muscle
Equipment:	• centrifuge with GSA-rotor • cellulose nitrate tubes • gauze • DEAE-sephadex column (2.5 x 60 cm)
Chemicals:	KCl, EDTA, MgCl$_2$, K/PO$_4$, ATP, DTT
Have ready:	**solution A:** 0.3 M KCl, 0.15 M potassium phosphate, 0.02 M EDTA, 0.005 M MgCl$_2$, 0.001 M ATP, pH 6.5
	solution B: 1.0 M KCl, 0.025 M EDTA, 0.06 M potassium phosphate, pH 6.5
	solution C: 0.6 M KCl, 0.025 M potassium phosphate, pH 6.5
phosphate,	**solution D:** 0.6 M KCl, 0.05 M potassium phosphate, pH 6.5
	solution E: 0.15 M potassium phosphate, 0.01 M EDTA, pH 7.5
	solution H: 0.04 M KCl, 0.01 M potassium phosphate, 0.001 M DTT, pH 6.5
	solution I: 3 M KCl, 0.01 M potassium phosphate, pH 6.5
Reference:	Margossian, S.S. and Lowey, S. (1982) Methods Enzymol. 85, 55-123

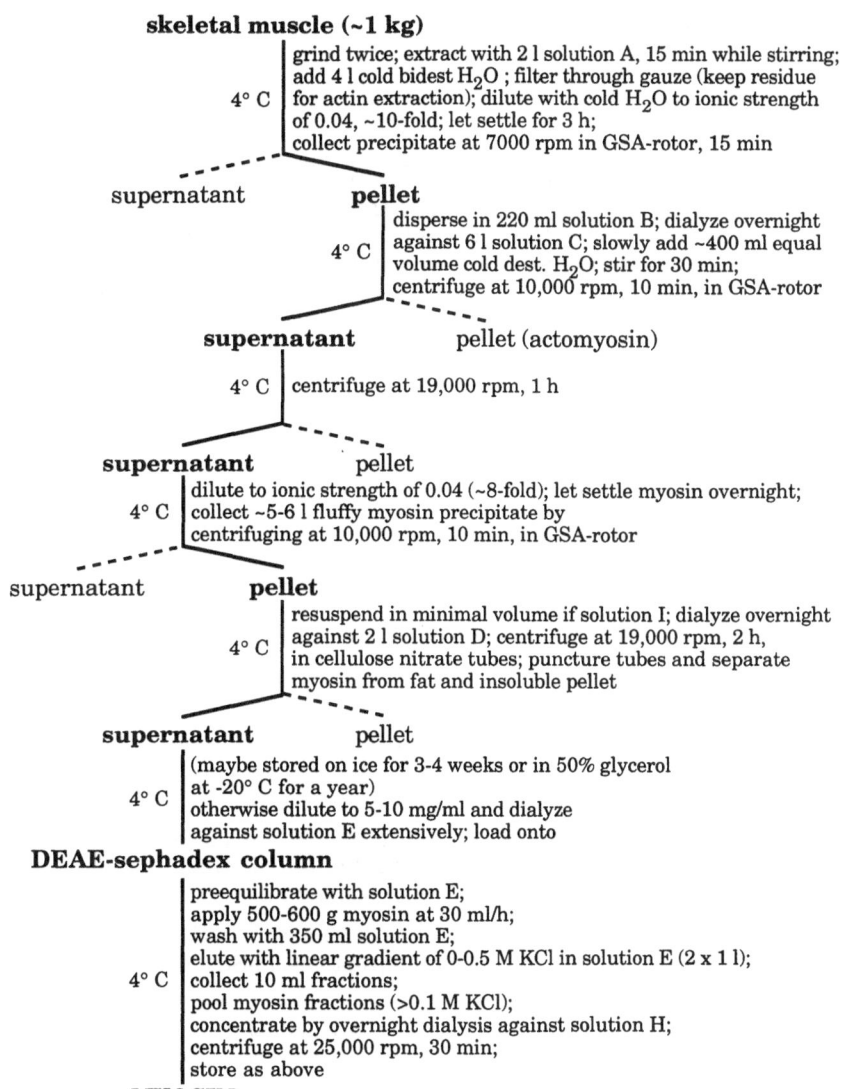

skeletal muscle (~1 kg)

4° C · grind twice; extract with 2 l solution A, 15 min while stirring; add 4 l cold bidest H_2O ; filter through gauze (keep residue for actin extraction); dilute with cold H_2O to ionic strength of 0.04, ~10-fold; let settle for 3 h; collect precipitate at 7000 rpm in GSA-rotor, 15 min

supernatant   **pellet**

4° C · disperse in 220 ml solution B; dialyze overnight against 6 l solution C; slowly add ~400 ml equal volume cold dest. H_2O; stir for 30 min; centrifuge at 10,000 rpm, 10 min, in GSA-rotor

supernatant   pellet (actomyosin)

4° C · centrifuge at 19,000 rpm, 1 h

supernatant   pellet

4° C · dilute to ionic strength of 0.04 (~8-fold); let settle myosin overnight; collect ~5-6 l fluffy myosin precipitate by centrifuging at 10,000 rpm, 10 min, in GSA-rotor

supernatant   **pellet**

4° C · resuspend in minimal volume if solution I; dialyze overnight against 2 l solution D; centrifuge at 19,000 rpm, 2 h, in cellulose nitrate tubes; puncture tubes and separate myosin from fat and insoluble pellet

supernatant   pellet

4° C · (maybe stored on ice for 3-4 weeks or in 50% glycerol at -20° C for a year) otherwise dilute to 5-10 mg/ml and dialyze against solution E extensively; load onto

DEAE-sephadex column

4° C · preequilibrate with solution E; apply 500-600 g myosin at 30 ml/h; wash with 350 ml solution E; elute with linear gradient of 0-0.5 M KCl in solution E (2 x 1 l); collect 10 ml fractions; pool myosin fractions (>0.1 M KCl); concentrate by overnight dialysis against solution H; centrifuge at 25,000 rpm, 30 min; store as above

MYOSIN

SMOOTH MUSCLE MYOSIN

A two-headed phosphorylatable motor enzyme consisting of two 200 kDa heavy chains and two pairs of 17 kDa and 20 kDa light chains. The α-helical coiled coil is 155 nm long. Phosphorylation of the 20 kDa light chains by Ca^{++}-calmodulin myosin light chain kinase increases actin-activated ATP-ase several hundred fold. Key mechanism for smooth muscle contraction.

Source: porcine aortic smooth muscle

Equipment:
- Polytron homogenizer
- centrifuge
- GSA-rotor
- Ti-45 rotor
- sepharose 4B gel filtration column (5 x 90 cm)
- water bath

Chemicals: NaCl, imidazole, DTT, KCl, MgSO$_4$, EGTA, EDTA, ATP, MOPS, streptomycin-SO$_4$, CaCl$_2$
protease inhibitors: PMSF

Have ready: **buffer A:** 0.15 M NaCl, 25 mM imidazole, 0.15 mM PMSF, pH 7.0 (12 l)

buffer B: 60 mM KCl, 20 mM imidazole, 1 mM MgSO$_4$, 1 mM DTT, 70 µM streptomycin-SO$_4$, 0.15 mM PMSF, pH 6.8 (500 ml)

buffer C: 0.6 M KCl, 20 mM imidazole, 2 mM EGTA, 1 mM EDTA, 1 mM DTT, 70 µM streptomycin-SO$_4$, 0.15 mM PMSF, pH 7.0 (250 ml)

buffer D: 35 mM KCl, 50 mM MOPS, 4 mM MgSO$_4$, 1 mM DTT, 70 µM streptomycin-SO$_4$, 0.15 mM
PMSF,   pH 7.0 (2 l)

buffer E: 0.5 M KCl, 0.1 mM MgSO$_4$, 2 mM EGTA, 0.1 mM imidazole, 1 mM DTT, 70 µM streptomycin-SO$_4$, 0.15 mM PMSF, pH 7.0 (5 l)

1 M MgSO$_4$ (25 ml), 5°C

ATP (10 ml) 50 mM ATP (pH 7.0) stored at -20°C

CaCl$_2$ (25 ml) 10 mM CaCl$_2$ stored at 5°C

EGTA (25 ml) 0.2 M EGTA (pH 7.0) stored at 5°C

Reference: Frederiksen, D.W. and Rees, D.D. (1982) Methods Enzymol. 85, 292-298
Gen-Bank: X06546 (chicken gizzard)

aortic media (~200 g)

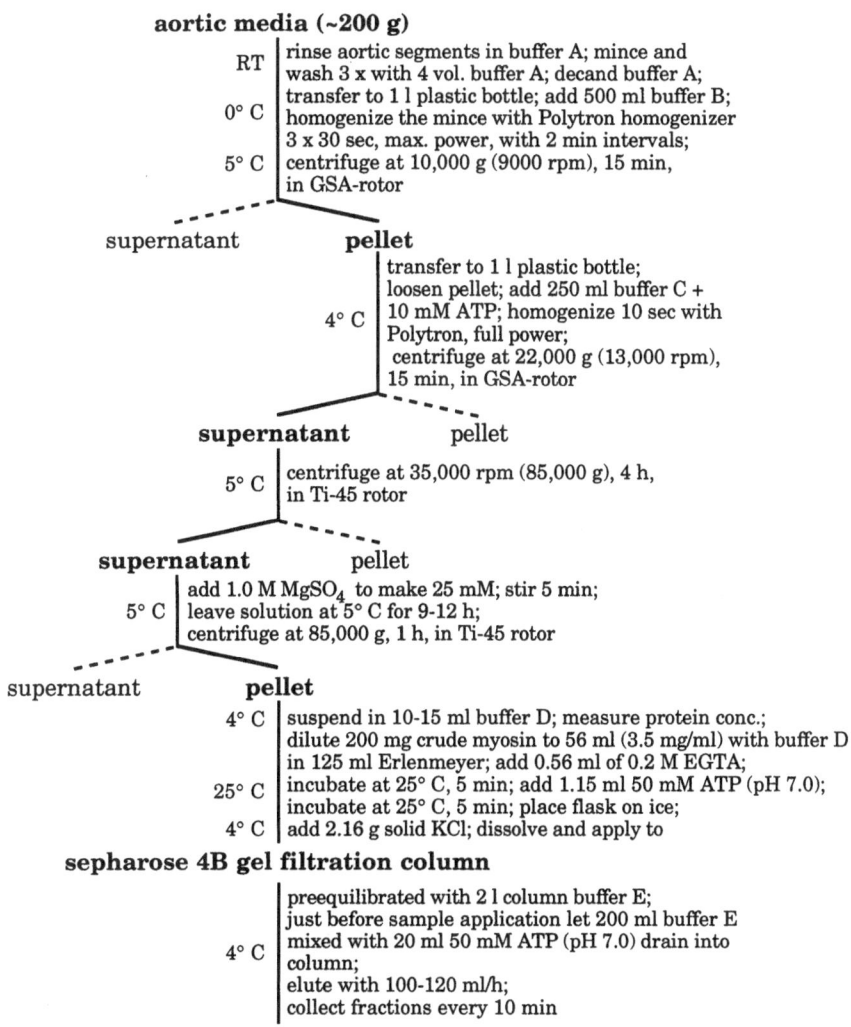

RT	rinse aortic segments in buffer A; mince and wash 3 x with 4 vol. buffer A; decand buffer A;
0° C	transfer to 1 l plastic bottle; add 500 ml buffer B; homogenize the mince with Polytron homogenizer 3 x 30 sec, max. power, with 2 min intervals;
5° C	centrifuge at 10,000 g (9000 rpm), 15 min, in GSA-rotor

supernatant   **pellet**

4° C	transfer to 1 l plastic bottle; loosen pellet; add 250 ml buffer C + 10 mM ATP; homogenize 10 sec with Polytron, full power; centrifuge at 22,000 g (13,000 rpm), 15 min, in GSA-rotor

supernatant   pellet

5° C	centrifuge at 35,000 rpm (85,000 g), 4 h, in Ti-45 rotor

supernatant   pellet

5° C	add 1.0 M $MgSO_4$ to make 25 mM; stir 5 min; leave solution at 5° C for 9-12 h; centrifuge at 85,000 g, 1 h, in Ti-45 rotor

supernatant   **pellet**

4° C	suspend in 10-15 ml buffer D; measure protein conc.; dilute 200 mg crude myosin to 56 ml (3.5 mg/ml) with buffer D in 125 ml Erlenmeyer; add 0.56 ml of 0.2 M EGTA;
25° C	incubate at 25° C, 5 min; add 1.15 ml 50 mM ATP (pH 7.0); incubate at 25° C, 5 min; place flask on ice;
4° C	add 2.16 g solid KCl; dissolve and apply to

sepharose 4B gel filtration column

4° C	preequilibrated with 2 l column buffer E; just before sample application let 200 ml buffer E mixed with 20 ml 50 mM ATP (pH 7.0) drain into column; elute with 100-120 ml/h; collect fractions every 10 min

SMOOTH MUSCLE MYOSIN
(80-120 mg, store at 5° C in buffer D for about a week)

IV Intermediate Filament Proteins

CYTOKERATINS

A multigene family of diverse 40-80 kDa type I (acidic) and type II (basic) polypeptides ranging from pI 4.8-8.0, constituting intermediate filaments (IF) in various epithelial cells and tissues.

Source: various cells and tissues

Equipment:
- Potter-Elvehjem
- Dounce homogenizer
- glass/teflon homogenizer
- gauze
- DEAE-cellulose (DE-52, Whatman) anion exchange column (2 x 6.5 cm)
- reversed phase HPLC (Pharmacia / LKB)
- Hi-Pore RP 304 (Bio Rad) column (250 x 4.6 mm) alternatively:
- μ CN column (μ Bondapak CN, Waters Ass.)

Chemicals: NaCl, KH_2PO_4, Na_2HPO_4, KCl, EDTA, DTT, Tris-HCl, Triton X-100, urea, 2-mercaptoethanol, guanidinium hydrochloride (u.p.), trifluoracetic acid (TFA), acetonitrile

Have ready: **buffer B:** 96 mM NaCl, 8 mM KH_2PO_4, 5.6 mM Na_2HPO_4, 1.5 mM KCl, 10 mM EDTA, 0.1 mM DTT, pH 6.8

high salt buffer: 2.0 M KCl, 200 mM NaCl, 10 mM Tris-HCl, pH 7.4 + 0.1 mM DTT

buffer D: 10 mM Tris-HCl, 140 mM NaCl, 1.5 M KCl, 5 mM EDTA, 0.5% Triton X-100, pH 7.6 + 5 mM DTT

PBS: 140 mM NaCl, 2.68 mM KCl, 8.1 mM Na_2HPO_4, 1.47 mM KH_2PO_4, pH 7.4

DEAE-sample buffer: 9.5 M urea, 10 mM Tris-HCl, 5% 2-mercaptoethanol, pH 8.0

DEAE-column buffer: 8 M urea, 30 mM Tris-HCl, 5 mM DTT, pH 8.0

Reference: Achtstaetter, T. et al. (1986) Methods Enzymol. 134, 355-371

tissue pieces
(from liver, kidney, epidermis, tongue)

4° C | mince with scalpel;
homogenize in >10 fold buffer B with glass/Teflon homogenizer;
filter through gauze; add 3 vol. high salt buffer; stir 30 min;
vigorously homogenize with Dounce;
centrifuge at 10,000 g, 20 min

- - - - -

supernatant   **pellet**

4° C | resuspend in equal vol. buffer D;
stir for 30 min;
centrifuge at 10,000 g, 30 min

- - - - -

pellet   **supernatant**

4° C | resuspend in and wash with PBS + 5 mM DTT;
centrifuge at 2500 g, 5 min;
dissolve in DEAE-sample buffer, 2 h;

25° C | centrifuge at 14,000 g, 15 min, 25° C;
dialyze overnight against 50 fold excess DEAE-column buffer;

25° C | centrifuge at 14,000 g, 15 min, 25° C;
measure protein conc.;

apply 50 mg protein to

DEAE-cellulose column

4° C | preequilibrate with DEAE-column buffer;
wash, elute with 20 ml/h with linear gradient of
2 x 200 ml 0-100 mM guanidinium hydrochloride;
collect 4 ml fractions;
pool individual peak fractions and apply to

reversed phase HPLC column

use 0.1% TFA and 0.07% TFA in acetonitrile
as aqueous and organic phase respectively;
equilibrate with 33% acetonitrile;
apply milliliters with 1-2 mg protein;
elute with 30 min linear gradient (33-48%)
acetonitrile at 2 ml/min;
screen by absorbance at 206 nm;
remove acetonitrile by vacuum evaporation;
lyophilize and store
(alternatively use μ Bondapak CN column)

CYTOKERATINS

DESMIN

A 53 kDa parallel homodimer with a central α-helical coiled-coil rod domain, a type III intermediate filament protein abundant and characteristic for smooth muscle.

Source:	chicken gizzard
Equipment:	• Polytron homogenizer • centrifuge, GSA-rotor • sephadex G-25 column (500 ml) • DEAE-cellulose (Whatman DE-52) column (2.5 x 10 cm)
Chemicals:	imidazole-HCl, KCl, EGTA, 2-mercaptoethanol, Triton X-100, KJ, DTT, ATP, urea, Tris-acetate, Na/PO$_4$
Have ready:	**buffer A:** 40 mM imidazole-HCl, 0.6 M KCl, 1 mM EGTA, 1 mM 2-mercaptoethanol, 0.5% Triton X-100, pH 6.9
	extraction buffer: 10 mM imidazole-HCl, 0.6 M KJ, 1 mM EGTA, 0.5% Triton X-100, 0.5 mM DTT, 0.2 mM ATP, pH 6.9
	urea buffer: 6 M urea, 10 mM sodium phosphate buffer, 5 mM EGTA, 0.1% 2-mercaptoethanol, pH 7.5
	0.01 M **Tris-acetate**, pH 8.2, 0.01 M 2-mercaptoethanol
Reference:	Geisler, N. and Weber, K. (1980) Europ. J. Biochem. 111, 425-433 Gen-Bank: NUC 21-8-89

208

chicken gizzard (~40 g)

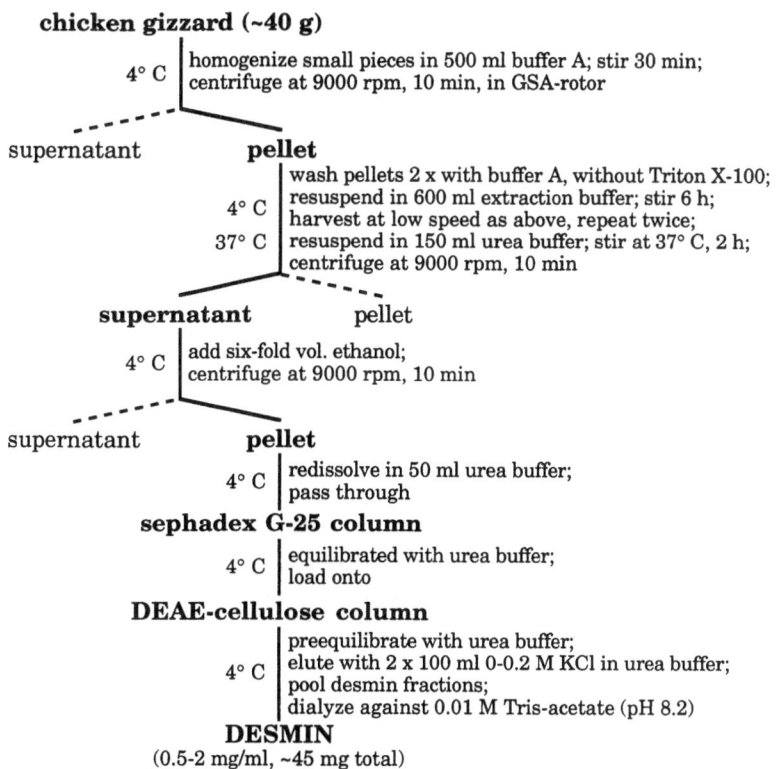

4° C | homogenize small pieces in 500 ml buffer A; stir 30 min;
centrifuge at 9000 rpm, 10 min, in GSA-rotor

supernatant   **pellet**

4° C | wash pellets 2 x with buffer A, without Triton X-100;
resuspend in 600 ml extraction buffer; stir 6 h;
harvest at low speed as above, repeat twice;
37° C | resuspend in 150 ml urea buffer; stir at 37° C, 2 h;
centrifuge at 9000 rpm, 10 min

supernatant   pellet

4° C | add six-fold vol. ethanol;
centrifuge at 9000 rpm, 10 min

supernatant   **pellet**

4° C | redissolve in 50 ml urea buffer;
pass through

sephadex G-25 column

4° C | equilibrated with urea buffer;
load onto

DEAE-cellulose column

4° C | preequilibrate with urea buffer;
elute with 2 x 100 ml 0-0.2 M KCl in urea buffer;
pool desmin fractions;
dialyze against 0.01 M Tris-acetate (pH 8.2)

DESMIN
(0.5-2 mg/ml, ~45 mg total)

DESMOCALMIN

> A 240 kDa, Ca⁺⁺-dependent calmodulin binding protein, constituative
> for desmosomes, probably a homodimer, consisting of ~100 nm long
> flexible rods, binding to keratin filaments.

Source: bovine muzzle

Equipment:
- centrifuge
- sepharose CL-4B (Pharmacia) column (3.5 x 50 cm)
- calmodulin-affinity column (Affi gel-calmodulin, Bio Rad)
- nylon mesh

Chemicals: sodium citrate, Nonidet P-40, EDTA, NaOH, KCl, DTT, Tris-HCl, EGTA, $CaCl_2$, $(NH_4)_2SO_4$
protease inhibitors: leupeptin, pepstatin A, PMSF

Have ready: **citrate buffer:** 0.1 M citric acid-sodium citrate buffer, pH 2.6 + 0.05% Nonidet P-40, 5 µg/ml leupeptin, 5 µg/ml pepstatin A

extraction buffer: 0.1 mM EDTA, 10 µg/ml leupeptin, pH 7.2

buffer A: 100 mM KCl, 20 mM Tris-HCl, 0.1 mM DTT, 1 µg/ml leupeptin, 1 µg/ml pepstatin A, 2 mM PMSF, pH 7.5

Reference: Tsukita, S. and Tsukita, S. (1985) J. Cell Biol. 101, 2070-2080

bovine muzzle (~7)

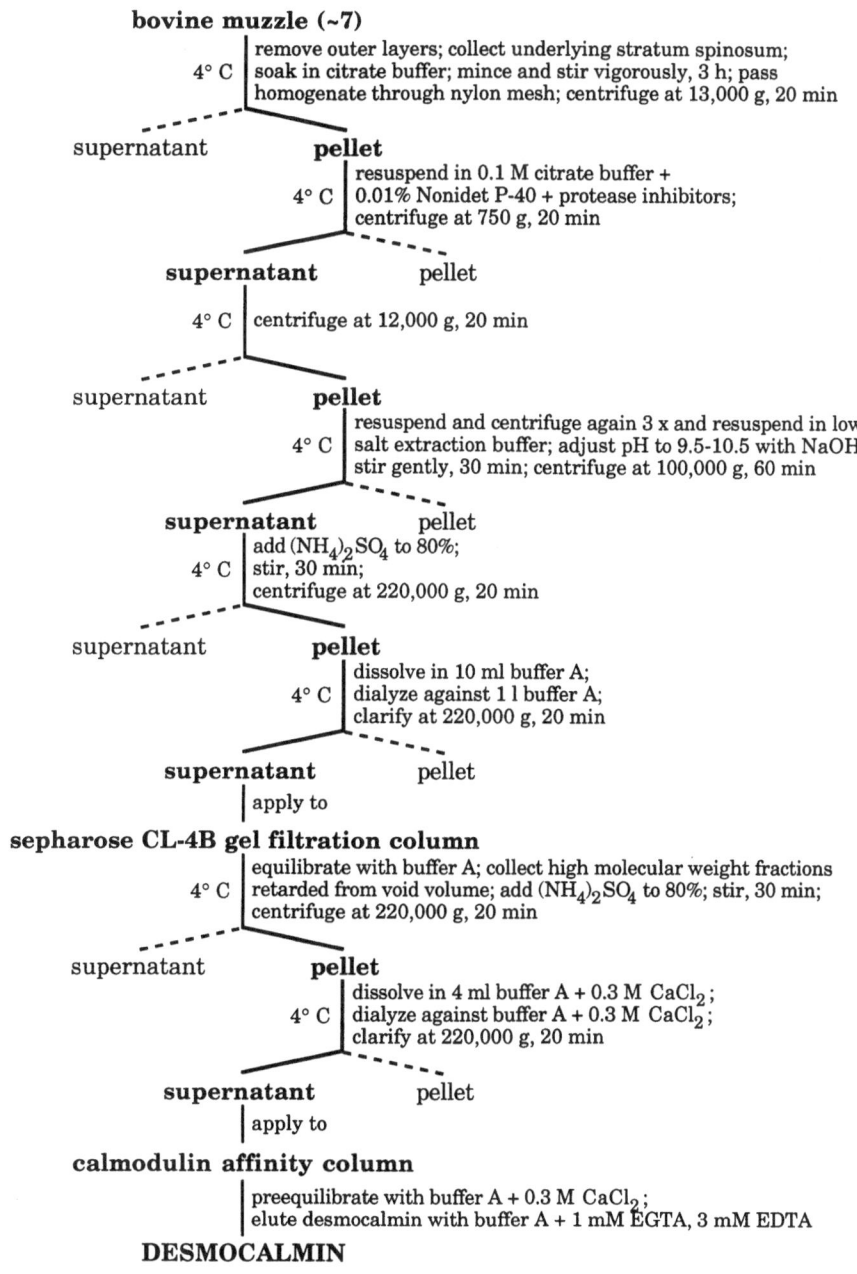

4° C │ remove outer layers; collect underlying stratum spinosum;
soak in citrate buffer; mince and stir vigorously, 3 h; pass
homogenate through nylon mesh; centrifuge at 13,000 g, 20 min

supernatant   **pellet**

4° C │ resuspend in 0.1 M citrate buffer +
0.01% Nonidet P-40 + protease inhibitors;
centrifuge at 750 g, 20 min

supernatant   pellet

4° C │ centrifuge at 12,000 g, 20 min

supernatant   **pellet**

4° C │ resuspend and centrifuge again 3 x and resuspend in low
salt extraction buffer; adjust pH to 9.5-10.5 with NaOH;
stir gently, 30 min; centrifuge at 100,000 g, 60 min

supernatant   pellet

4° C │ add $(NH_4)_2SO_4$ to 80%;
stir, 30 min;
centrifuge at 220,000 g, 20 min

supernatant   **pellet**

4° C │ dissolve in 10 ml buffer A;
dialyze against 1 l buffer A;
clarify at 220,000 g, 20 min

supernatant   pellet

│ apply to

sepharose CL-4B gel filtration column

4° C │ equilibrate with buffer A; collect high molecular weight fractions
retarded from void volume; add $(NH_4)_2SO_4$ to 80%; stir, 30 min;
centrifuge at 220,000 g, 20 min

supernatant   **pellet**

4° C │ dissolve in 4 ml buffer A + 0.3 M $CaCl_2$;
dialyze against buffer A + 0.3 M $CaCl_2$;
clarify at 220,000 g, 20 min

supernatant   pellet

│ apply to

calmodulin affinity column

│ preequilibrate with buffer A + 0.3 M $CaCl_2$;
elute desmocalmin with buffer A + 1 mM EGTA, 3 mM EDTA

DESMOCALMIN

FILAGGRINS

Mouse filaggrin, a 26 kDa intermediate filament (IF) associated protein with the capacity to promote keratin filament assembly.

Source:	mouse epidermis
Equipment:	• Polytron homogenizer • DEAE-cellulose column • SDS-preparative gel electrophoresis
Chemicals:	Tris-HCl, urea, 2-mercaptoethanol, DTT **protease inhibitors:** PMSF
Have ready:	**PBS:** phosphate buffered saline + 1 mM PMSF
	extraction buffer: 8 M urea, 0.1 M Tris-HCl, pH 7.5, 0.1 M 2-mercaptoethanol, 0.5 mM PMSF
	DEAE-column buffer: 8 M urea, 10 mM Tris-HCl, pH 7.4, 1 mM DTT, 0.5 mM PMSF
Reference:	Steinert, P.M. et al. (1981) Proc. Natl. Acad. Sci. 78, 4097-4101 Gen-Bank: J02929 (human), J05198 (mouse)

stratum corneum
of new born mice epidermis

0-4° C	remove, wash 2 x in PBS + 1 mM PMSF; extract with extraction buffer, 1 h; homogenize with Polytron; extract 2 h; load onto

DEAE-cellulose column

4° C	preequilibrate and wash with DEAE-column buffer; recover filaggrin in the flow through fractions; further purify by preparative SDS-gel electrophoresis

FILAGGRIN
(~20 mg/g dry weight of tissue)

FILENSIN

Source: porcine eye lens

Equipment:
- centrifuge
- Waring blender
- DEAE-cellulose column
- hydroxylapatide column

Chemicals: Tris-HCl, KCl, $MgCl_2$, DTT, EDTA, urea, Na/PO_4, **protease inhibitors:** PMSF, leupeptin

Have ready: **TKM-buffer:** 50 mM Tris-HCl, 25 mM KCl, 5 mM $MgCl_2$, pH 8.0 + 1 mM DTT, 1 mM PMSF, 2 µg/ml leupeptin

extraction buffer: 7 M urea, 10 mM Tris-HCl, 2 mM EDTA, 1 mM DTT, 0.5 mM PMSF, pH 7.6

urea-buffer: 7-8 M urea, 10 mM Tris-HCl, pH 7.3, 1 mM PMSF

phosphate-buffer: 7 M urea, 10 mM Na/PO_4, pH 7.6, 1 mM DTT, 0.5 mM PMSF

Reference: Merdes, A. et al. (1991) J. Cell Biol. 115, 397-410

214

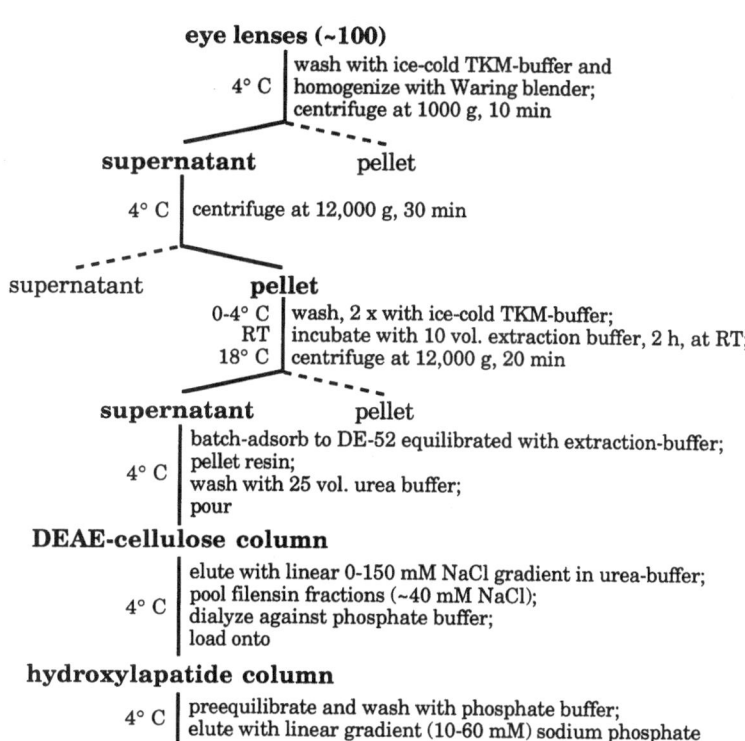

eye lenses (~100)

4° C | wash with ice-cold TKM-buffer and homogenize with Waring blender; centrifuge at 1000 g, 10 min

supernatant   pellet

4° C | centrifuge at 12,000 g, 30 min

supernatant   **pellet**

0-4° C | wash, 2 x with ice-cold TKM-buffer;
RT | incubate with 10 vol. extraction buffer, 2 h, at RT;
18° C | centrifuge at 12,000 g, 20 min

supernatant   pellet

4° C | batch-adsorb to DE-52 equilibrated with extraction-buffer; pellet resin; wash with 25 vol. urea buffer; pour

DEAE-cellulose column

4° C | elute with linear 0-150 mM NaCl gradient in urea-buffer; pool filensin fractions (~40 mM NaCl); dialyze against phosphate buffer; load onto

hydroxylapatide column

4° C | preequilibrate and wash with phosphate buffer; elute with linear gradient (10-60 mM) sodium phosphate

FILENSIN

GLIAL FIBRILLARY ACIDIC PROTEIN (GFAP)

A 49.8 kDa type III intermediate filament protein, 48 nm in length, forming antiparallel overlapping dimers with 64 nm repeats. Present in cells of glial origin.

Source: porcine brain

Equipment:
- Dounce homogenizer
- centrifuge, SW-27 rotors (Beckman Instr.)
- hydroxylapatide column (1.5 x 10 cm)
- water bath
- DEAE-cellulose (Whatman DE-52) column (2.5 x 20 cm)
- CM-cellulose column (2 x 20 cm)
- vacuum dialysis equipment

Chemicals: sucrose, Na/PO_4, NaCl, Triton X-100, urea, EGTA, EDTA, Tris-HCl, 2-mercaptoethanol, sodium formate
protease inhibitors: PMSF

Have ready: **solution A:** 10 mM phosphate buffer, pH 6.8, 1 mM EDTA, 0.1 M NaCl

10 mM phosphate buffer, pH 7.4, 8 M deionized urea, 1% 2-mercaptoethanol

deionized urea: prepare by chromatography on mixed ion exchange column (Bio Rad AE501-X8) to remove cyanates

dialyzing buffer: 10 mM Tris-HCl, pH 8.3, 1 mM EDTA, 2 mM PMSF

buffer B: 20 mM Tris-HCl, pH 7.5, 50 mM 2-mercaptoethanol, 2 mM EGTA, 2 mM PMSF, 8 M urea

DEAE-column buffer: 20 mM Tris-HCl, pH 7.5, 50 mM 2-mercaptoethanol, 2 mM EGTA, 1 mM PMSF, 7 M urea

buffer C: 20 mM sodium formate buffer, pH 4.0, 50 mM 2-mercaptoethanol, 2 mM EGTA, 1 mM PMSF, 7 M urea

Reference: Liem, R. (1982) J. Neurochem. 38, 142-150
Inagaki, M. (1990) J. Biol. Chem. 265, 4722-4729
Gen-Bank: K01347 (mouse), J04569 (human)

porcine brain (white matter), ~175 g

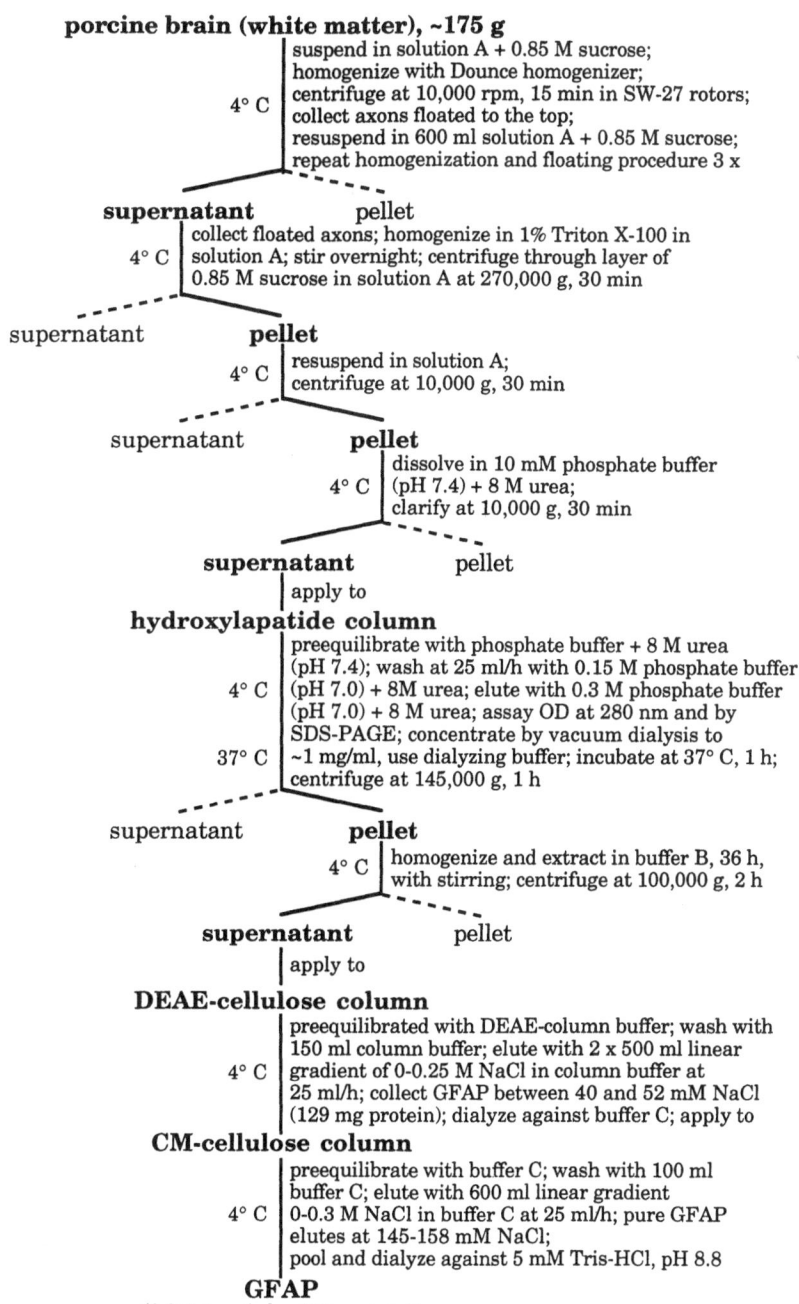

	suspend in solution A + 0.85 M sucrose;
	homogenize with Dounce homogenizer;
4° C	centrifuge at 10,000 rpm, 15 min in SW-27 rotors;
	collect axons floated to the top;
	resuspend in 600 ml solution A + 0.85 M sucrose;
	repeat homogenization and floating procedure 3 x

supernatant   pellet

collect floated axons; homogenize in 1% Triton X-100 in
4° C solution A; stir overnight; centrifuge through layer of
0.85 M sucrose in solution A at 270,000 g, 30 min

supernatant   **pellet**

4° C resuspend in solution A;
centrifuge at 10,000 g, 30 min

supernatant   **pellet**

dissolve in 10 mM phosphate buffer
4° C (pH 7.4) + 8 M urea;
clarify at 10,000 g, 30 min

supernatant   pellet

apply to

hydroxylapatide column

preequilibrate with phosphate buffer + 8 M urea
(pH 7.4); wash at 25 ml/h with 0.15 M phosphate buffer
4° C (pH 7.0) + 8M urea; elute with 0.3 M phosphate buffer
(pH 7.0) + 8 M urea; assay OD at 280 nm and by
SDS-PAGE; concentrate by vacuum dialysis to
37° C ~1 mg/ml, use dialyzing buffer; incubate at 37° C, 1 h;
centrifuge at 145,000 g, 1 h

supernatant   **pellet**

4° C homogenize and extract in buffer B, 36 h,
with stirring; centrifuge at 100,000 g, 2 h

supernatant   pellet

apply to

DEAE-cellulose column

preequilibrated with DEAE-column buffer; wash with
150 ml column buffer; elute with 2 x 500 ml linear
4° C gradient of 0-0.25 M NaCl in column buffer at
25 ml/h; collect GFAP between 40 and 52 mM NaCl
(129 mg protein); dialyze against buffer C; apply to

CM-cellulose column

preequilibrate with buffer C; wash with 100 ml
buffer C; elute with 600 ml linear gradient
4° C 0-0.3 M NaCl in buffer C at 25 ml/h; pure GFAP
elutes at 145-158 mM NaCl;
pool and dialyze against 5 mM Tris-HCl, pH 8.8

GFAP

(0.3-0.8 mg/ml, ~100 mg total)

217

α-INTERNEXIN

A type IV intermediate filament subunit with an apparent MW of 68 kDa but with a calculated MW of 55 kDa, primarily distributed in the central nervous system, capable of self assembly in vitro.

Source: rat spinal cord and optic nerve

Equipment:
- centrifuge
- SW 27 rotor, SW 41 rotor (Beckman Instr.)
- hydroxylapatide (Bio Rad) column (1.5 x 10 cm)
- DEAE-cellulose (Whatman DE-52) column (0.75 x 5 cm)
- vacuum concentration device

Chemicals: Na/PO$_4$, EDTA, NaCl, Triton X-100, sucrose, 2-mercaptoethanol, urea, KCl, EGTA, MgCl$_2$, ATP
protease inhibitors: PMSF

Have ready: **solution A:** 10 mM phosphate buffer, pH 6.8, 1 mM EDTA, 0.1 M NaCl

solubilization buffer: 10 mM phosphate buffer, pH 7.4, 1% 2-mercaptoethanol, 8 M deionized urea

assembly buffer: 10 mM phosphate, pH 6.8, 0.1 M KCl, 1 mM EGTA, 1 mM MgCl$_2$, 1 mM ATP, 1 mM PMSF

Reference: Pachter, J.S. and Liem, R.K.H. (1985) J. Cell Biol. 101, 1316-1322
Gen-Bank: M73049 (rat brain)

rat spinal cord and optic nerve

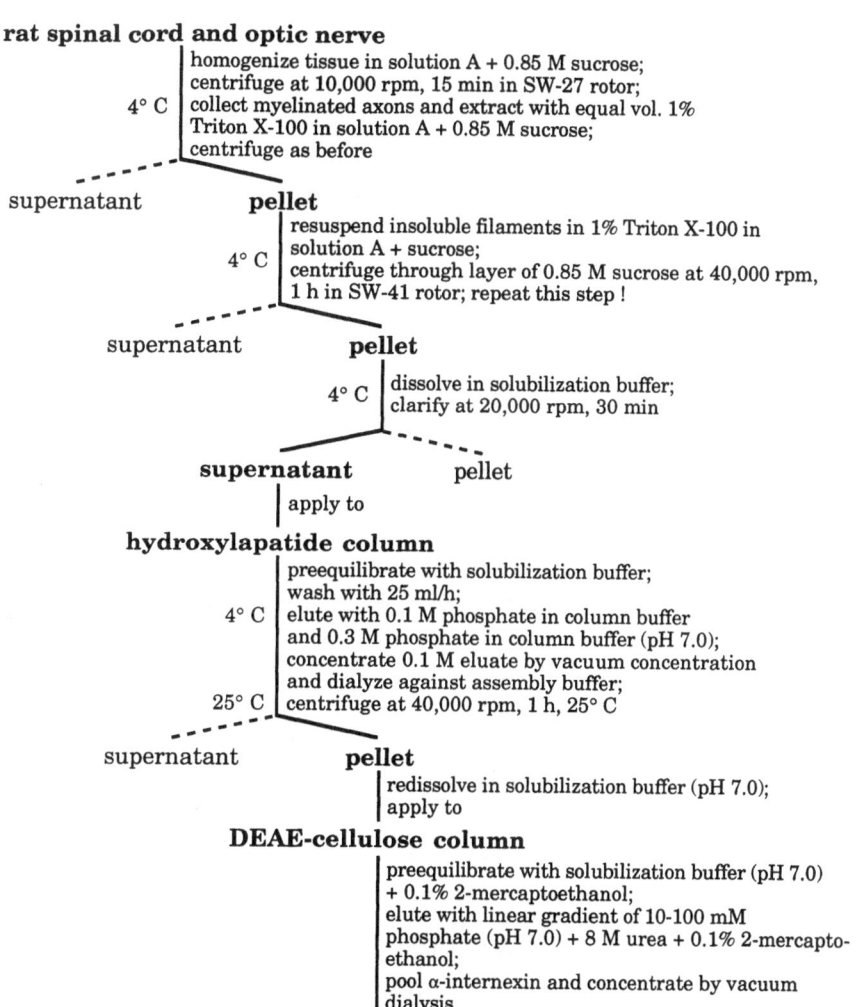

4° C — homogenize tissue in solution A + 0.85 M sucrose;
centrifuge at 10,000 rpm, 15 min in SW-27 rotor;
collect myelinated axons and extract with equal vol. 1%
Triton X-100 in solution A + 0.85 M sucrose;
centrifuge as before

supernatant   **pellet**

4° C — resuspend insoluble filaments in 1% Triton X-100 in
solution A + sucrose;
centrifuge through layer of 0.85 M sucrose at 40,000 rpm,
1 h in SW-41 rotor; repeat this step !

supernatant   **pellet**

4° C — dissolve in solubilization buffer;
clarify at 20,000 rpm, 30 min

supernatant   pellet

apply to

hydroxylapatide column

4° C — preequilibrate with solubilization buffer;
wash with 25 ml/h;
elute with 0.1 M phosphate in column buffer
and 0.3 M phosphate in column buffer (pH 7.0);
concentrate 0.1 M eluate by vacuum concentration
and dialyze against assembly buffer;

25° C — centrifuge at 40,000 rpm, 1 h, 25° C

supernatant   **pellet**

redissolve in solubilization buffer (pH 7.0);
apply to

DEAE-cellulose column

preequilibrate with solubilization buffer (pH 7.0)
+ 0.1% 2-mercaptoethanol;
elute with linear gradient of 10-100 mM
phosphate (pH 7.0) + 8 M urea + 0.1% 2-mercapto-
ethanol;
pool α-internexin and concentrate by vacuum
dialysis

α-INTERNEXIN

LAMINS

A family of 60-75 kDa intermediate filament type proteins constituating the nuclear lamina. α-type lamins include lamins A and C (pI~7), β-type lamins include lamins B (pI~6).

Source: rat liver

Equipment:
- Potter-Elvehjem
- cheese cloth
- centrifuge, SW-28 rotor (Beckman Instr.)
- phosphocellulose (Whatman P-11) column (0.5 ml)
- DEAE-cellulose (Whatman DE-52) column (0.5 ml)

Chemicals: Tris-HCl, KCl, $MgCl_2$, triethanolamine-HCl,
sucrose, DTT, DNA-se I, RNA-se A, Triton X-100,
MES-KOH, EDTA, glycine, urea
protease inhibitors: PMSF

Have ready: **TKM-buffer:** 0.05 M Tris-HCl, pH 7.5, 0.025 M KCl, 0.005M $MgCl_2$

0.1 M $MgCl_2$ + 0.0005 M PMSF + 0.001 M DTT

100 μg/ml **DNA-se I** + 1 mg/ml **RNA-se A**

10% sucrose solution: 10% sucrose, 10 mM triethanolamine-HCl, pH 8.5, 0.1 mM $MgCl_2$

30% sucrose solution: 30% sucrose, 10 mM triethanolamine-HCl, 0.1 mM $MgCl_2$, pH 7.5

extraction solution I: 10% sucrose, 2% Triton X-100, 20 mM MES-KOH, 300 mM KCl, 2 mM EDTA, 1 mM DTT, pH 6.0

extraction solution II: 2% Triton X-100, 20 mM Tris-HCl, 500 mM KCl, 2 mM EDTA, 1 mM DTT, pH 9.0

urea-buffer: 6 M urea, 20 mM Tris-HCl, pH 9.0

buffer U1: 6 M urea, 20 mM Tris-HCl, pH 9.0, 500 mM KCl, 2 mM EDTA, 1 mM DTT

buffer U2: 8.5 M urea, 10 mM MES, pH 6.5, 10 mM glycine, 0.5 mM EDTA, 1 mM DTT

Reference: Aebi, U. et al. (1986) Nature 323, 560-564

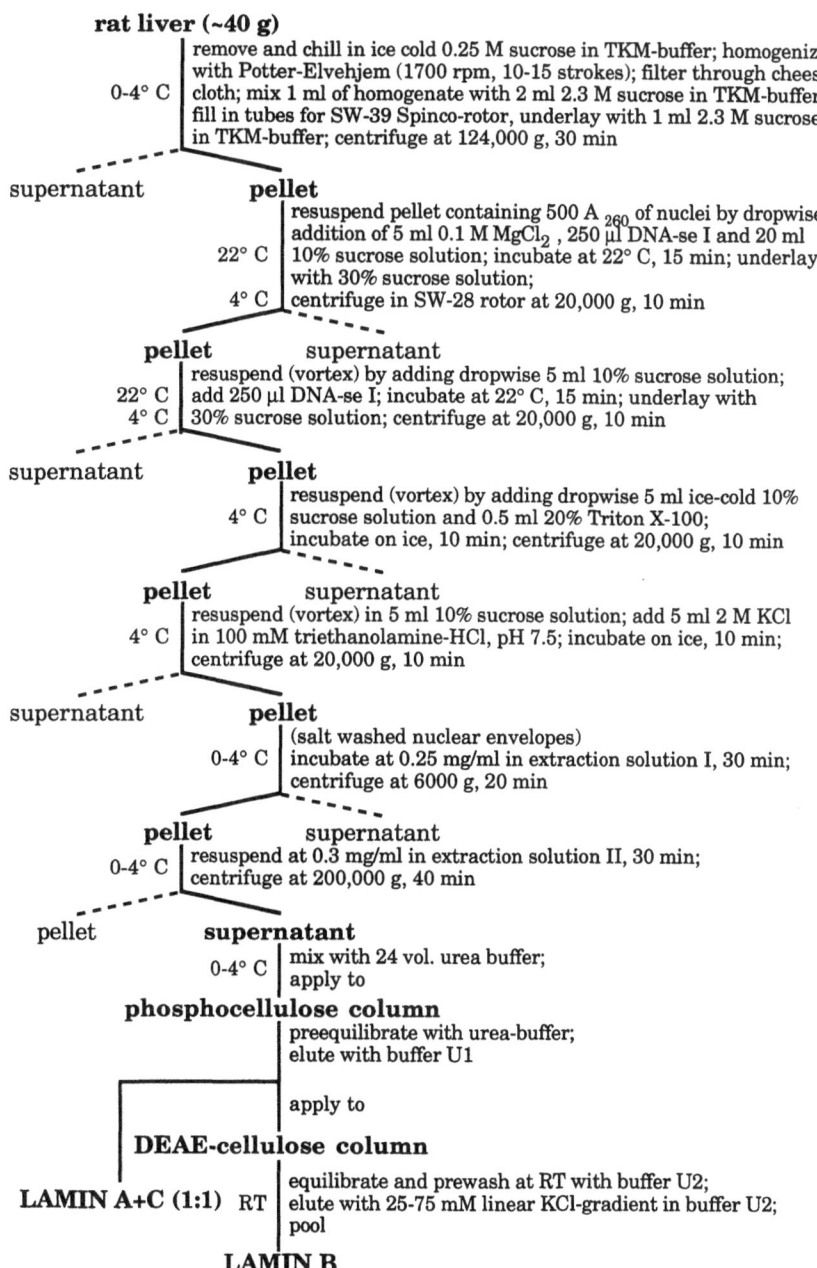

rat liver (~40 g)

0-4° C | remove and chill in ice cold 0.25 M sucrose in TKM-buffer; homogenize with Potter-Elvehjem (1700 rpm, 10-15 strokes); filter through cheese cloth; mix 1 ml of homogenate with 2 ml 2.3 M sucrose in TKM-buffer; fill in tubes for SW-39 Spinco-rotor, underlay with 1 ml 2.3 M sucrose in TKM-buffer; centrifuge at 124,000 g, 30 min

supernatant — **pellet**

22° C
4° C | resuspend pellet containing 500 A$_{260}$ of nuclei by dropwise addition of 5 ml 0.1 M MgCl$_2$, 250 µl DNA-se I and 20 ml 10% sucrose solution; incubate at 22° C, 15 min; underlay with 30% sucrose solution; centrifuge in SW-28 rotor at 20,000 g, 10 min

pellet — supernatant

22° C
4° C | resuspend (vortex) by adding dropwise 5 ml 10% sucrose solution; add 250 µl DNA-se I; incubate at 22° C, 15 min; underlay with 30% sucrose solution; centrifuge at 20,000 g, 10 min

supernatant — **pellet**

4° C | resuspend (vortex) by adding dropwise 5 ml ice-cold 10% sucrose solution and 0.5 ml 20% Triton X-100; incubate on ice, 10 min; centrifuge at 20,000 g, 10 min

pellet — supernatant

4° C | resuspend (vortex) in 5 ml 10% sucrose solution; add 5 ml 2 M KCl in 100 mM triethanolamine-HCl, pH 7.5; incubate on ice, 10 min; centrifuge at 20,000 g, 10 min

supernatant — **pellet**

0-4° C | (salt washed nuclear envelopes) incubate at 0.25 mg/ml in extraction solution I, 30 min; centrifuge at 6000 g, 20 min

pellet — supernatant

0-4° C | resuspend at 0.3 mg/ml in extraction solution II, 30 min; centrifuge at 200,000 g, 40 min

pellet — **supernatant**

0-4° C | mix with 24 vol. urea buffer; apply to

phosphocellulose column

| preequilibrate with urea-buffer; elute with buffer U1

apply to

DEAE-cellulose column

LAMIN A+C (1:1)   RT | equilibrate and prewash at RT with buffer U2; elute with 25-75 mM linear KCl-gradient in buffer U2; pool

LAMIN B

NEUROFILAMENT TRIPLET PROTEIN

Polypeptides of 210 kDa (NF-H), 160 kDa (NF-M) and 68 kDa (NF-L), the major components of neuronal intermediate filaments (NFs).

Source:	porcine spinal cord
Equipment:	• centrifuge • sephadex G-25 column • DEAE-cellulose (Whatman DE-52) column (50 ml)
Chemicals:	MES, EGTA, MgCl$_2$, glycerol, Na/PO$_4$, urea, 2-mercaptoethanol
Have ready:	**homogenization buffer:** 0.1 M MES, 1 mM EGTA, 0.5 mM MgCl$_2$, pH 6.5 **buffer A:** 10 mM sodium phosphate buffer, pH 7.5, 1 mM EGTA, 0.1% 2-mercaptoethanol, 6 M urea

Reference:	Geisler, N. and Weber, K. (1981) J. Mol. Biol. 151, 665-571
	Gen-Bank:
	      X15306, X15309 (human NF-H)
	      Y00067 (human NF-M)
	      X05608 (human NF-L)
	      M23349, M24494-M24496 (murine NF-H)
	      X05640 (murine NF-M)
	      M13016 (murine NF-L)

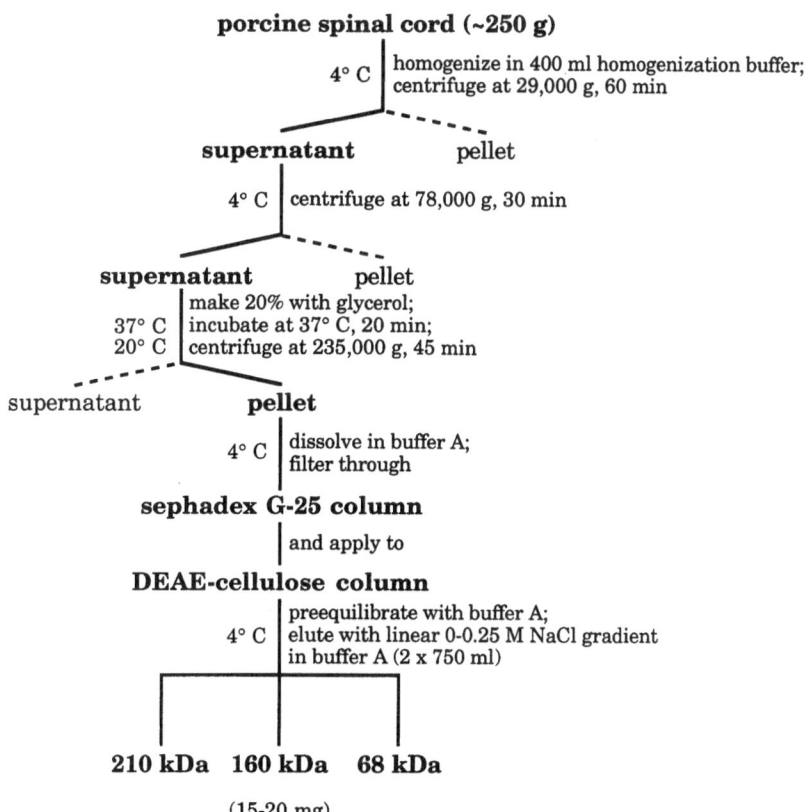

porcine spinal cord (~250 g)

4° C | homogenize in 400 ml homogenization buffer;
centrifuge at 29,000 g, 60 min

supernatant   pellet

4° C | centrifuge at 78,000 g, 30 min

supernatant   pellet

37° C | make 20% with glycerol;
20° C | incubate at 37° C, 20 min;
centrifuge at 235,000 g, 45 min

supernatant   **pellet**

4° C | dissolve in buffer A;
filter through

sephadex G-25 column

| and apply to

DEAE-cellulose column

4° C | preequilibrate with buffer A;
elute with linear 0-0.25 M NaCl gradient
in buffer A (2 x 750 ml)

210 kDa   160 kDa   68 kDa

(15-20 mg)

PLECTIN

A 466 kDa protein running at 300 kDa on SDS-PAGE. Molecules composed of 190 nm long rods flanked by two globes of 4 nm diameter on each side. Ubiquitous, and a versatile crosslinking protein for intermediate filament matrices.

Source: bovine eye lenses

Equipment:
- centrifuge
- Dounce homogenizer
- teflon/glass homogenizer
- sephacryl S-500 (Pharmacia) column (2.6 x 50 cm)
- SDS-PAGE, 6.25%

Chemicals: Na/PO$_4$, NaCl, KCl, MgCl$_2$, Triton X-100, DNA-se I, RNA-se A, 2-mercaptoethanol, EGTA, sodium N-lauroyl-sarcosinate (SLS), liquid nitrogen
protease inhibitors: PMSF

Have ready: **homogenization buffer:** 6 mM sodium phosphate buffer, pH 7.1, 771 mM NaCl, 3 mM KCl, 100 mM MgCl$_2$, 8 mM 2-mercaptoethanol, 2 mM PMSF, 1% Triton X-100 + 50 µg/ml DNA-se I, 20 µg/ml ribonuclease A

extraction solution II: 0.3 mM sodium phosphate, pH 8.9, 2 mM PMSF, 8 mM 2-mercaptoethanol, 0.1 mM EGTA, 1% sodium N-lauroyl sarcosinate (SLS)

solution A: 0.3 mM sodium phosphate, pH 8.7, 5 mM EGTA, 1 mM PMSF, 0.25% SLS

Reference: Weitzer, G. and Wiche, G. (1987) Eur. J. Biochem. 169, 41-52
Gen-Bank: X59601 (rat)

100 eye lenses (~200 g)

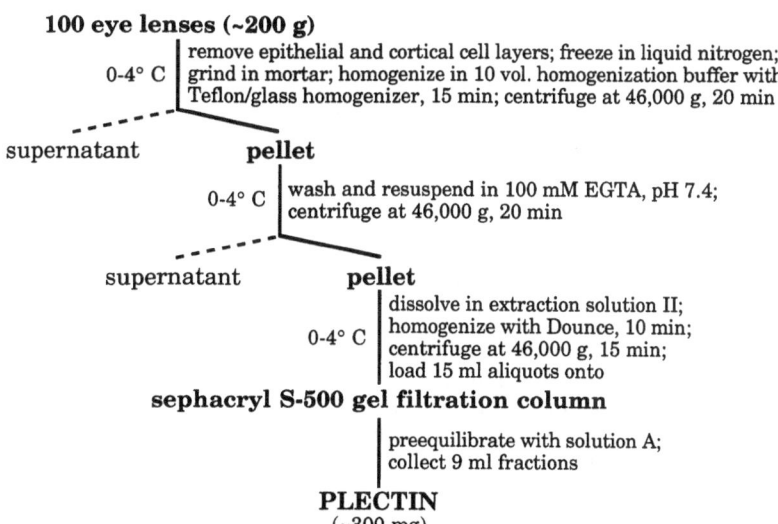

0-4° C | remove epithelial and cortical cell layers; freeze in liquid nitrogen; grind in mortar; homogenize in 10 vol. homogenization buffer with Teflon/glass homogenizer, 15 min; centrifuge at 46,000 g, 20 min

supernatant   **pellet**

0-4° C | wash and resuspend in 100 mM EGTA, pH 7.4; centrifuge at 46,000 g, 20 min

supernatant   **pellet**

0-4° C | dissolve in extraction solution II; homogenize with Dounce, 10 min; centrifuge at 46,000 g, 15 min; load 15 ml aliquots onto

sephacryl S-500 gel filtration column

preequilibrate with solution A; collect 9 ml fractions

PLECTIN
(~300 mg)

PLECTIN

A phosphorylatable globular tetramer of 980 kDa with an apparent MW of 230 kDa on SDS-gels, associated with desmin and vimentin intermediate filaments.

Source: chicken gizzard

Equipment:
- centrifuge, siliconized glass tubes
- Lourdes homogenizer
- Waring blender
- hydroxylapatide (BioRad) column (100 ml bed. vol.)
- DEAE-sephacel column (in siliconized glass) (45 ml)
- phosphocellulose column (in siliconized glass) (20 ml)

Chemicals: Tris-HCl, EGTA, KCl, 2-mercaptoethanol, KJ, $Na_2S_2O_3$, acetone, urea, DTT, K/PO_4, NaCl
protease inhibitors: PMSF, TAME

Have ready: **buffer A:** 10 mM Tris-HCl, 140 mM KCl, 2 mM EGTA, 0.1% 2-mercaptoethanol, 0.2 mM PMSF, pH 7.5

low salt buffer: 10 mM Tris-HCl, 10 mM EGTA, 0.1% 2-mercaptoethanol, 0.2 mM PMSF, pH 7.5

high salt buffer: 10 mM Tris-HCl, 0.6 M KJ, 10 mM $Na_2S_2O_3$, 0.1% 2-mercaptoethanol, 0.2 mM PMSF, pH 7.5

cold acetone

buffer B: 10 mM Tris-HCl, 6 M urea, 2 mM DTT, 5 mM TAME, 0.2 mM PMSF, pH 7.5

K-phosphate buffer: 10-65 mM K/PO_4 + 6 M urea, 2 mM DTT, 5 mM TAME, pH 7.5

dialysis buffer I: 20 mM Tris-HCl, 6 M urea, 2 mM EGTA, 2 mM DTT, 1 mM TAME, pH 7.5

dialysis buffer II: 10 mM Tris-HCl, 6 M urea, 2 mM EGTA, 2 mM DTT, 1 mM TAME, pH 7.0

Reference: Sandoval, I.V. et al. (1983) J. Biol. Chem. 258, 2568-2576

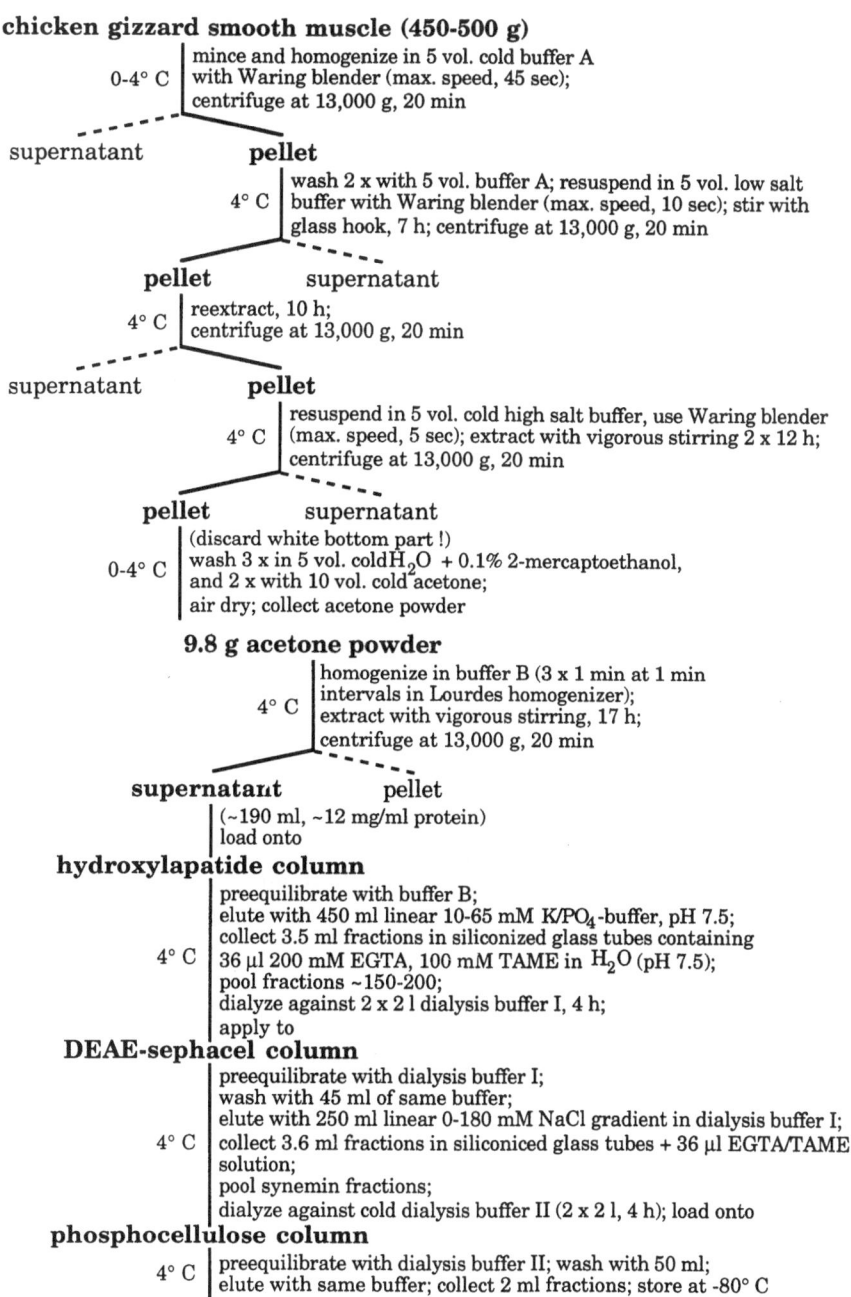

chicken gizzard smooth muscle (450-500 g)

0-4° C | mince and homogenize in 5 vol. cold buffer A with Waring blender (max. speed, 45 sec); centrifuge at 13,000 g, 20 min

supernatant   **pellet**

4° C | wash 2 x with 5 vol. buffer A; resuspend in 5 vol. low salt buffer with Waring blender (max. speed, 10 sec); stir with glass hook, 7 h; centrifuge at 13,000 g, 20 min

pellet   supernatant

4° C | reextract, 10 h; centrifuge at 13,000 g, 20 min

supernatant   **pellet**

4° C | resuspend in 5 vol. cold high salt buffer, use Waring blender (max. speed, 5 sec); extract with vigorous stirring 2 x 12 h; centrifuge at 13,000 g, 20 min

pellet   supernatant

0-4° C | (discard white bottom part !) wash 3 x in 5 vol. coldH_2O + 0.1% 2-mercaptoethanol, and 2 x with 10 vol. cold acetone; air dry; collect acetone powder

9.8 g acetone powder

4° C | homogenize in buffer B (3 x 1 min at 1 min intervals in Lourdes homogenizer); extract with vigorous stirring, 17 h; centrifuge at 13,000 g, 20 min

supernatant   pellet

(~190 ml, ~12 mg/ml protein) load onto

hydroxylapatide column

4° C | preequilibrate with buffer B; elute with 450 ml linear 10-65 mM K/PO_4-buffer, pH 7.5; collect 3.5 ml fractions in siliconized glass tubes containing 36 μl 200 mM EGTA, 100 mM TAME in H_2O (pH 7.5); pool fractions ~150-200; dialyze against 2 x 2 l dialysis buffer I, 4 h; apply to

DEAE-sephacel column

4° C | preequilibrate with dialysis buffer I; wash with 45 ml of same buffer; elute with 250 ml linear 0-180 mM NaCl gradient in dialysis buffer I; collect 3.6 ml fractions in siliconiced glass tubes + 36 μl EGTA/TAME solution; pool synemin fractions; dialyze against cold dialysis buffer II (2 x 2 l, 4 h); load onto

phosphocellulose column

4° C | preequilibrate with dialysis buffer II; wash with 50 ml; elute with same buffer; collect 2 ml fractions; store at -80° C

SYNEMIN

VIMENTIN

A 54 kDa type III intermediate filament protein with various isoforms through phosphorylation, typically for mesenchymally derived cells and tissues.

Source: pig eye lenses

Equipment:
- centrifuge
- Polytron Potter
- sephadex G-25 column (250 ml bed vol.)
- carboxymethyl-cellulose (Whatman CM-32) column (50 ml)
- sephadex G-25 column (50 ml)
- DEAE-cellulose (Whatman DE-52) column (10 ml bed vol.)
- SDS-PAGE

Chemicals: Tris-HCl, $MgCl_2$, 2-mercaptoethanol, DTT, urea, EGTA, sodium formate, NaCl, ethanol

Have ready: **homogenization buffer:** 50 mM Tris-HCl, 5 mM $MgCl_2$, 10 mM 2-mercaptoethanol, pH 7.4

extraction buffer: 50 mM Tris-HCl, pH 7.4, 5 mM $MgCl_2$, 8 M urea, 5 mM DTT

formate buffer: 30 mM sodium formate, pH 4.0, 1 mM EGTA, 1 mM DTT, 7 M urea

DEAE-column buffer: 7 M urea, 50 mM Tris-HCl, 1 mM EGTA, 1 mM DTT, 10 mM NaCl, pH 7.5

Reference: Geisler, N. and Weber, K. (1981) FEBS-Lett. 125, 253-256

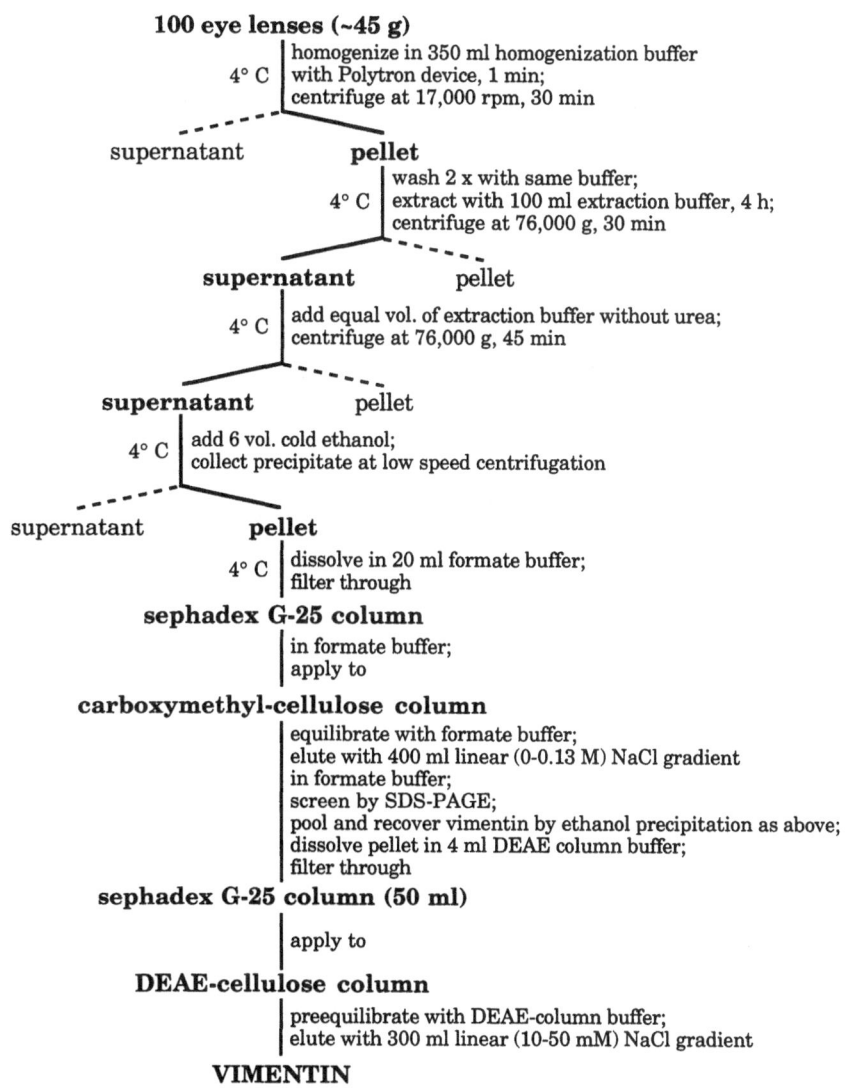

100 eye lenses (~45 g)

4° C | homogenize in 350 ml homogenization buffer
      | with Polytron device, 1 min;
      | centrifuge at 17,000 rpm, 30 min

supernatant   **pellet**

4° C | wash 2 x with same buffer;
      | extract with 100 ml extraction buffer, 4 h;
      | centrifuge at 76,000 g, 30 min

supernatant   pellet

4° C | add equal vol. of extraction buffer without urea;
      | centrifuge at 76,000 g, 45 min

supernatant   pellet

4° C | add 6 vol. cold ethanol;
      | collect precipitate at low speed centrifugation

supernatant   **pellet**

4° C | dissolve in 20 ml formate buffer;
      | filter through

sephadex G-25 column

| in formate buffer;
| apply to

carboxymethyl-cellulose column

| equilibrate with formate buffer;
| elute with 400 ml linear (0-0.13 M) NaCl gradient
| in formate buffer;
| screen by SDS-PAGE;
| pool and recover vimentin by ethanol precipitation as above;
| dissolve pellet in 4 ml DEAE column buffer;
| filter through

sephadex G-25 column (50 ml)

| apply to

DEAE-cellulose column

| preequilibrate with DEAE-column buffer;
| elute with 300 ml linear (10-50 mM) NaCl gradient

VIMENTIN
(10-20 mg)

V Plasmamembrane and Organelle-Associated Cytoskeletal Proteins

ADDUCIN

A heterodimeric calmodulin binding protein from the erythrocyte membrane skeleton composed of α (103 kDa) and β (97 kDa) subunits, also present in vertebrate brain.

Source: erythrocytes

Equipment:
- centrifuge
- Leuko-Pak filter
- Millipore Pellicon Cassette + Durapore filter (0.5 μm)
- DEAE-cellulose (DE-53, Whatman) column (12 ml)
- linear sucrose gradient mixer
- Ti-45 rotor, VTi-50 rotor (Beckman Instr.)

Chemicals: Na/PO_4, NaCl, Dextran T-500, NaEGTA, DTT, NaEDTA, Triton X-100, PEG 8000, sucrose, HEPES
protease inhibitors: DFP, PMSF, pepstatin A, leupeptin

Have ready: **washing buffer I:** 150 mM NaCl, 5 mM sodium phosphate, 0.75% Dextran T-500, pH 7.5

washing buffer II: 10 mM sodium phosphate, 150 mM NaCl

washing buffer III: 7.5 mM sodium phosphate, 1 mM NaEGTA, 1 mM DTT, 0.01% DFP

extraction solution I: 10 mM sodium phosphate, 100 mM NaCl, 1 mM NaEDTA, 0.5% Triton X-100, 200 μg/ml PMSF, 4 μg/ml pepstatin A, 4 μg/ml leupeptin, pH 7.5

extraction solution II: 1 M NaCl, 1 mM NaEDTA, 10 mM sodium phosphate, 1 mM DTT, 200 μg/ml PMSF, 4 μg/ml pepstatin A, 4 μg/ml leupeptin, pH 7.5

dialyzing buffer: 10 mM sodium phospate, 50 mM NaCl, 0.25 mM NaEGTA, 1 mM DTT, pH 7.5

sucrose gradient buffer: 10 mM HEPES, 100 mM NaCl, 0.25 mM NaEGTA, 1 mM DTT, pH 7.5

Reference: Gardner, K. and Bennett, V. (1986) J. Biol. Chem 261, 1339-1348
Gen-Bank: X58141 (human α-adducin), X58199 (human β-adducin)

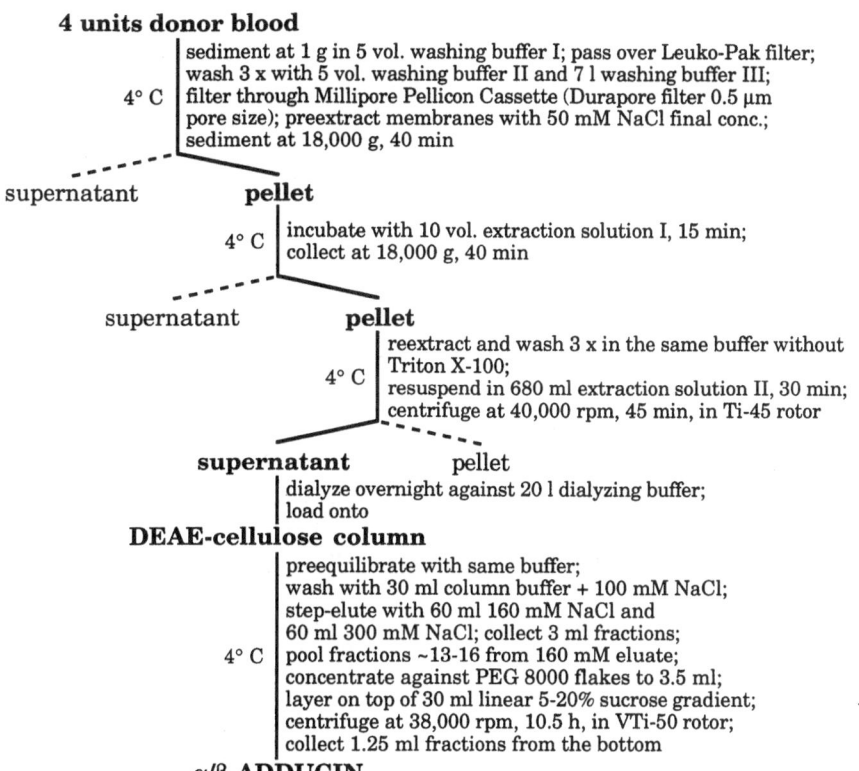

4 units donor blood

4° C | sediment at 1 g in 5 vol. washing buffer I; pass over Leuko-Pak filter; wash 3 x with 5 vol. washing buffer II and 7 l washing buffer III; filter through Millipore Pellicon Cassette (Durapore filter 0.5 μm pore size); preextract membranes with 50 mM NaCl final conc.; sediment at 18,000 g, 40 min

supernatant   **pellet**

4° C | incubate with 10 vol. extraction solution I, 15 min; collect at 18,000 g, 40 min

supernatant   **pellet**

4° C | reextract and wash 3 x in the same buffer without Triton X-100; resuspend in 680 ml extraction solution II, 30 min; centrifuge at 40,000 rpm, 45 min, in Ti-45 rotor

supernatant   pellet

dialyze overnight against 20 l dialyzing buffer; load onto

DEAE-cellulose column

4° C | preequilibrate with same buffer; wash with 30 ml column buffer + 100 mM NaCl; step-elute with 60 ml 160 mM NaCl and 60 ml 300 mM NaCl; collect 3 ml fractions; pool fractions ~13-16 from 160 mM eluate; concentrate against PEG 8000 flakes to 3.5 ml; layer on top of 30 ml linear 5-20% sucrose gradient; centrifuge at 38,000 rpm, 10.5 h, in VTi-50 rotor; collect 1.25 ml fractions from the bottom

α/β ADDUCIN
(3-5 mg)

233

ANKYRIN

A family of 202-206 kDa proteins, asymetric and monomeric in solution, responsible for coupling the actin-spectrin network to plasma membranes.

Source: pig brain

Equipment:
- Polytron homogenizer (large head)
- centrifuge, Ti-45, Ti-60 rotors
- sephacryl S-500 (Pharmacia) column (5 x 85 cm)
- spectrin-CNBr-activated sepharose 4B affinity column (1.5 x 14 cm)
- hydroxylapatide column (0.7 x 5 cm)

Chemicals: sucrose, NaEGTA, NaN_3, KCl, Na/PO_4, Tween-20, KI, DTT, NaCl, NaBr, $(NH_4)_2SO_4$

protease inhibitors: DFP, PMSF, leupeptin, pepstatin A

Have ready:

solution I: 0.32 M sucrose, 2 mM NaEGTA, 1 mM NaN_3, 5 µg/ml leupeptin, 5 µg/ml pepstatin A, 200 µg/ml PMSF, 0.01% DFP, pH 7.5

solution II: 0.8 M sucrose, 2 mM NaEGTA, 50 µg/ml PMSF, pH 7.5

phosphate buffer: 10 mM Na/PO_4, 2 mM NaEGTA, 1 mM NaN_3, 50 µg/ml PMSF, pH 7.5

extraction solution: 0.8 M KI, 0.05% Tween-20, 0.5 mM DTT in phosphate buffer

dialysis buffer: 0.2 M NaCl, 10 mM Na/PO_4, 2 mM NaEGTA, 1 mM NaN_3, 0.05% Tween-20, 0.5 mM DTT, pH 7.5

S-500 buffer: 1 M NaBr, 10 mM Na/PO_4, 2 mM NaEGTA, 0.05% Tween-20, 0.5 mM DTT, 1 mM NaN_3, pH 7.5

spectrin affinity column elution buffer: 0.8 M KI, 10 mM Na/PO_4, 0.05% Tween-20, 1 mM NaN_3, 0.5 mM DTT, pH 7.5

HT-column buffer: 1 M NaBr, 10 mM Na/PO_4, 0.05% Tween-20, 1 mM NaN_3, 0.5 mM DTT, pH 7.5

Reference: Davis, J.Q. and Bennett, V. (1984) J. Biol. Chem. 259, 13550-13559
Gen-Bank: X16609, X56958, X56957 (human)

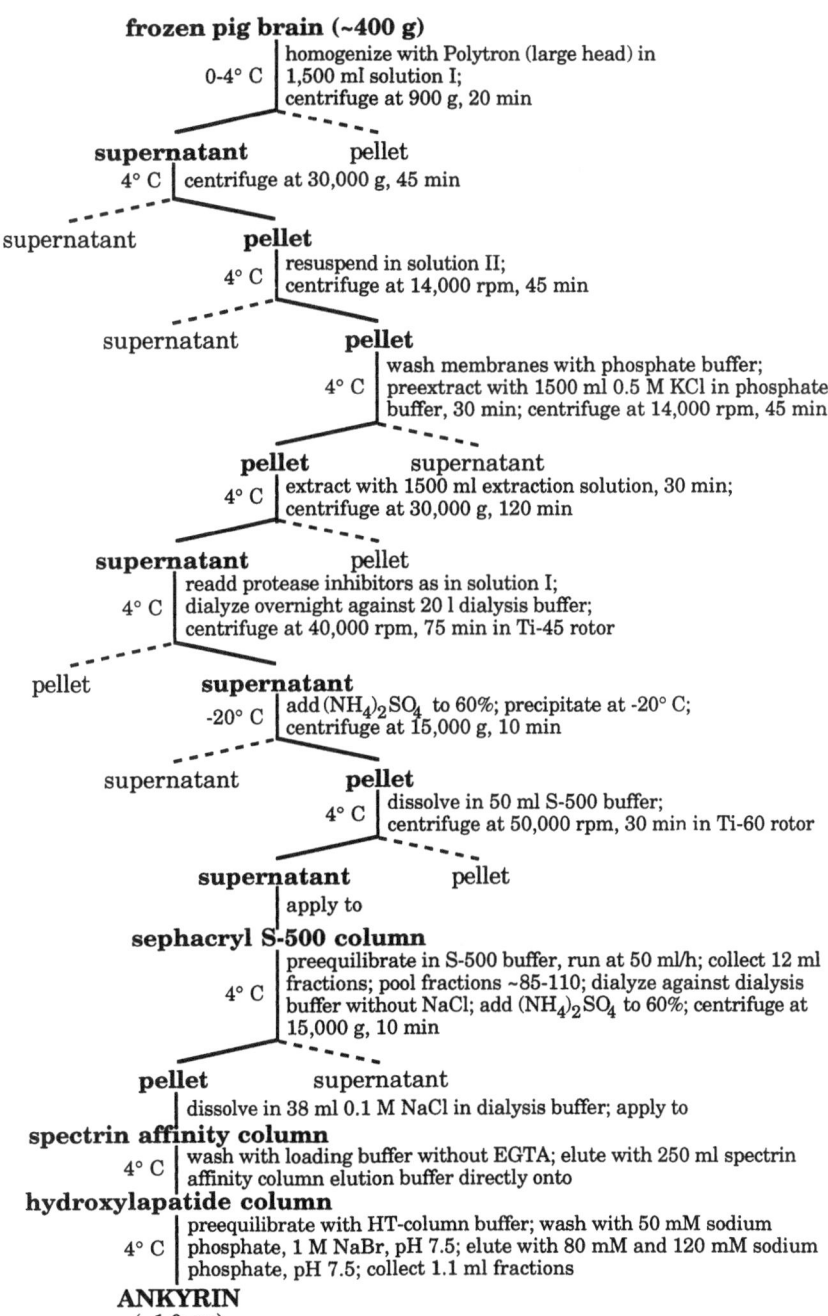

frozen pig brain (~400 g)

0-4° C | homogenize with Polytron (large head) in 1,500 ml solution I; centrifuge at 900 g, 20 min

supernatant   pellet

4° C | centrifuge at 30,000 g, 45 min

supernatant   **pellet**

4° C | resuspend in solution II; centrifuge at 14,000 rpm, 45 min

supernatant   **pellet**

4° C | wash membranes with phosphate buffer; preextract with 1500 ml 0.5 M KCl in phosphate buffer, 30 min; centrifuge at 14,000 rpm, 45 min

pellet   supernatant

4° C | extract with 1500 ml extraction solution, 30 min; centrifuge at 30,000 g, 120 min

supernatant   pellet

4° C | readd protease inhibitors as in solution I; dialyze overnight against 20 l dialysis buffer; centrifuge at 40,000 rpm, 75 min in Ti-45 rotor

pellet   **supernatant**

-20° C | add $(NH_4)_2SO_4$ to 60%; precipitate at -20° C; centrifuge at 15,000 g, 10 min

supernatant   **pellet**

4° C | dissolve in 50 ml S-500 buffer; centrifuge at 50,000 rpm, 30 min in Ti-60 rotor

supernatant   pellet

| apply to

sephacryl S-500 column

4° C | preequilibrate in S-500 buffer, run at 50 ml/h; collect 12 ml fractions; pool fractions ~85-110; dialyze against dialysis buffer without NaCl; add $(NH_4)_2SO_4$ to 60%; centrifuge at 15,000 g, 10 min

pellet   supernatant

| dissolve in 38 ml 0.1 M NaCl in dialysis buffer; apply to

spectrin affinity column

4° C | wash with loading buffer without EGTA; elute with 250 ml spectrin affinity column elution buffer directly onto

hydroxylapatide column

4° C | preequilibrate with HT-column buffer; wash with 50 mM sodium phosphate, 1 M NaBr, pH 7.5; elute with 80 mM and 120 mM sodium phosphate, pH 7.5; collect 1.1 ml fractions

ANKYRIN
(~1.0 mg)

AP-180

A monomeric, clathrin associated protein with an apparent MW of 180 kDa on SDS-gels but with native molecular weight of 119 kDa. A promotor for clathrin assembly into coat structures.

Source:	bovine brain
Equipment:	• centrifuge
	• Dounce homogenizer
	• Eppendorf microfuge
	• Waring blender
	• Centricon-30 microconcentrator
	• superose-6 (Pharmacia) gel filtration column (10 x 300 mm)
	• Mono Q HR 16/10 (Pharmacia) anion exchange column
	• hydroxylapatide/FPLC column (5 x 50 mm)
Chemicals:	MES, EGTA, $MgCl_2$, NaN_3, Ficoll, sucrose, Tris-HCl, EDTA, DTT, ethanolamine, K/PO_4
	protease inhibitors: PMSF
Have ready:	**homogenization buffer:** 0.1 M MES, 1 mM EGTA, 0.5 mM $MgCl_2$, 0.02% NaN_3, pH 6.5
	extraction buffer: 0.5 M Tris-HCl, 50 mM MES, 0.25 mM $MgCl_2$, 0.5 mM EGTA, 1 mM EDTA, 1 mM DTT, 0.1 mM PMSF, pH 7.0
	superose-6 buffer: 0.5 M Tris-HCl, 2 mM EDTA, 1 mM DTT, 0.02% NaN_3, pH 7.0
	dialyzing buffer: 20 mM ethanolamine, pH 9.0, 2 mM EDTA, 1 mM DTT
	HT-buffer: 0.5 M Tris-HCl, 2 mM K/PO_4, pH 7.0
Reference:	Ahle, S. and Ungewickell, E. (1986) EMBO J. 5, 3143-3149

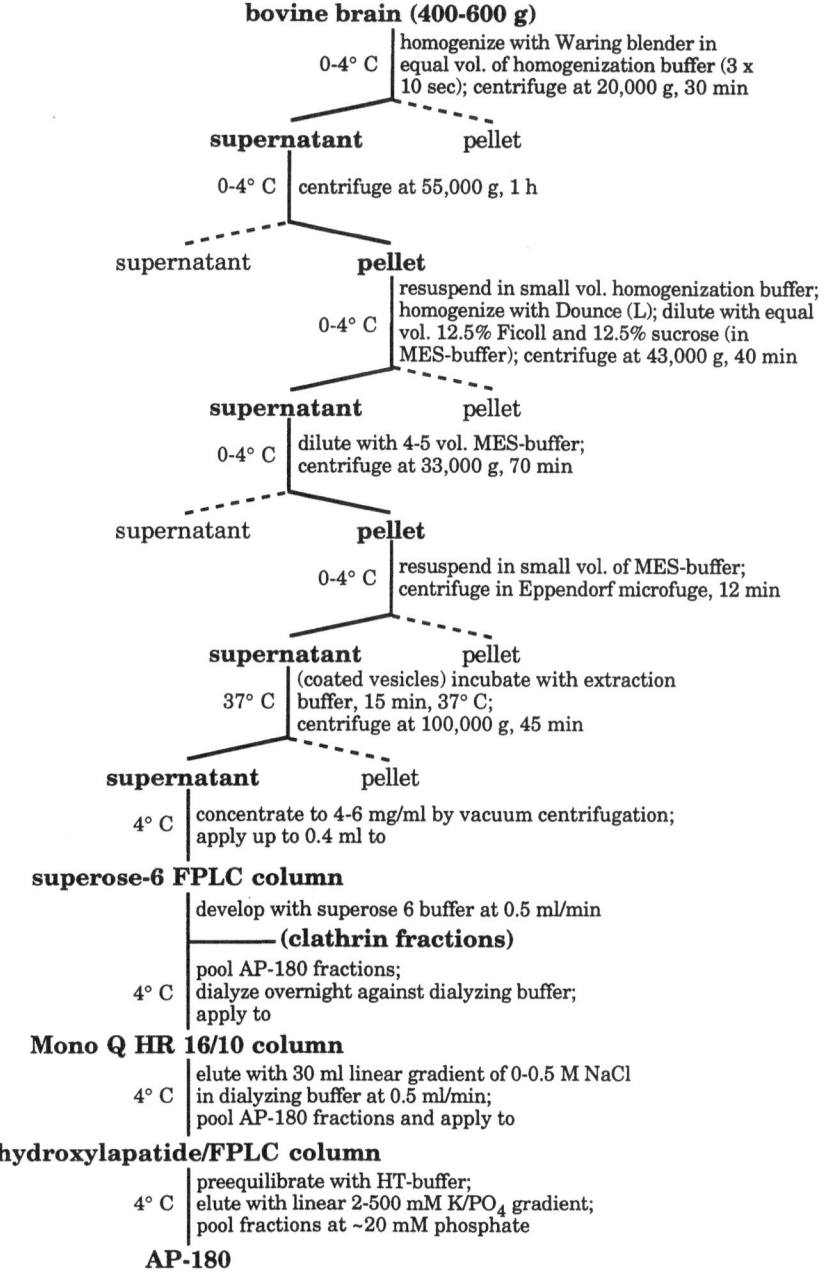

bovine brain (400-600 g)

0-4° C | homogenize with Waring blender in equal vol. of homogenization buffer (3 x 10 sec); centrifuge at 20,000 g, 30 min

supernatant   pellet

0-4° C | centrifuge at 55,000 g, 1 h

supernatant   **pellet**

0-4° C | resuspend in small vol. homogenization buffer; homogenize with Dounce (L); dilute with equal vol. 12.5% Ficoll and 12.5% sucrose (in MES-buffer); centrifuge at 43,000 g, 40 min

supernatant   pellet

0-4° C | dilute with 4-5 vol. MES-buffer; centrifuge at 33,000 g, 70 min

supernatant   **pellet**

0-4° C | resuspend in small vol. of MES-buffer; centrifuge in Eppendorf microfuge, 12 min

supernatant   pellet

37° C | (coated vesicles) incubate with extraction buffer, 15 min, 37° C; centrifuge at 100,000 g, 45 min

supernatant   pellet

4° C | concentrate to 4-6 mg/ml by vacuum centrifugation; apply up to 0.4 ml to

superose-6 FPLC column

develop with superose 6 buffer at 0.5 ml/min

———(clathrin fractions)

4° C | pool AP-180 fractions; dialyze overnight against dialyzing buffer; apply to

Mono Q HR 16/10 column

4° C | elute with 30 ml linear gradient of 0-0.5 M NaCl in dialyzing buffer at 0.5 ml/min; pool AP-180 fractions and apply to

hydroxylapatide/FPLC column

4° C | preequilibrate with HT-buffer; elute with linear 2-500 mM K/PO$_4$ gradient; pool fractions at ~20 mM phosphate

AP-180

AUXILIN

> **A 86 kDa single copy clathrin associated protein, promoting assembly of clathrin triskelia into coat structures, running at 110 kDa in SDS-gels.**

Source: bovine brain

Equipment:
- Waring blender
- water bath
- Centricon-30 microconcentrator
- superose-6 (Pharmacia) column (10 x 30 mm)
- hydroxylapatide/FPLC column (5 x 50 mm)
- equipment for AP-180 purification

Chemicals: MES, EGTA, $MgCl_2$, NaN_3, Ficoll, sucrose, Tris-HCl, EDTA, DTT, ethanolamine, K/PO_4
protease inhibitors: PMSF

Have ready: **homogenization buffer:** 0.1 M MES, 1 mM EGTA, 0.5 mM $MgCl_2$, 0.02% NaN_3, pH 6.5

extraction buffer: 0.5 M Tris-HCl, 50 mM MES, 0.25 mM $MgCl_2$, 0.5 mM EGTA, 1 mM EDTA, 1 mM DTT, 0.1 mM PMSF, pH 7.0

superose-6 buffer: 0.5 M Tris-HCl, 2 mM EDTA, 1 mM DTT, 0.02% NaN_3, pH 7.0

dialyzing buffer: 20 mM ethanolamine, pH 9.0, 2 mM EDTA, 1 mM DTT

HT-buffer: 0.5 M Tris-HCl, 2 mM K/PO_4, pH 7.0

Reference: Ahle, S. and Ungewickell, E. (1990) J. Cell Biol. 111, 19-29

bovine brain
(follow AP-180 purification, using 60 mg coat
protein for extraction with 0.5 M Tris, pH 7.0)

| fractionate ~12 mg assembly proteins on

hydroxylapatide/FPLC column

| pool auxilin fractions eluting between
| 0.1-0.15 M phosphate;
| concentrate by vacuum microconcentration;
| load 3 batches onto

superose-6 gel filtration column

| preequilibrate and develop with
| superose-6 buffer

AUXILIN
(~200 µg)

BAND 6 POLYPEPTIDE

A 75 kDa protein associated with cytoplasmic desmosomal plaque structures.

Source:	bovine muzzle epidermis
Equipment:	• centrifuge • Polytron homogenizer • Branson sonifier • water bath • DEAE-cellulose (Whatman DE-52) column (2 x 6.5 cm) • Mono-S cation exchange column (Pharmacia)
Chemicals:	citric acid, NP-40, DTT, Tris-HCl, urea, DTE, guanidinium-HCl **protease inhibitors:** L-3-trans-carboxy-oxiran-2-carbonyl-1-leugmatin ("E64", Peptide Inst. Osaka, Japan), PMSF, leupeptin, pepstatin
Have ready:	**CASC A1 buffer:** 0.1 M citric acid, 0.05% NP-40, 1 mM DTT, 1 mM PMSF, 0.5 mM "E64", 5 µg/ml leupeptin, 5 µg/ml pepstatin, pH 2.3
	CASC B1 buffer: 0.1 M citric acid, 0.01% NP-40, 1 mM DTT, protease inhibitors as above, pH 2.3
	extraction buffer: 5 mM Tris-buffer, pH 8.5, 9.0 M urea, 5 mM DTT, 1 mM PMSF
	DEAE-column buffer: 8 M urea, 30 mM Tris-HCl, 2.5 mM DTE, 1 mM PMSF, pH 8.5
	phosphate buffered saline (PBS)
Reference:	Kapprell, H.P. et al. (1988) J. Cell Biol. 106, 1679-1691

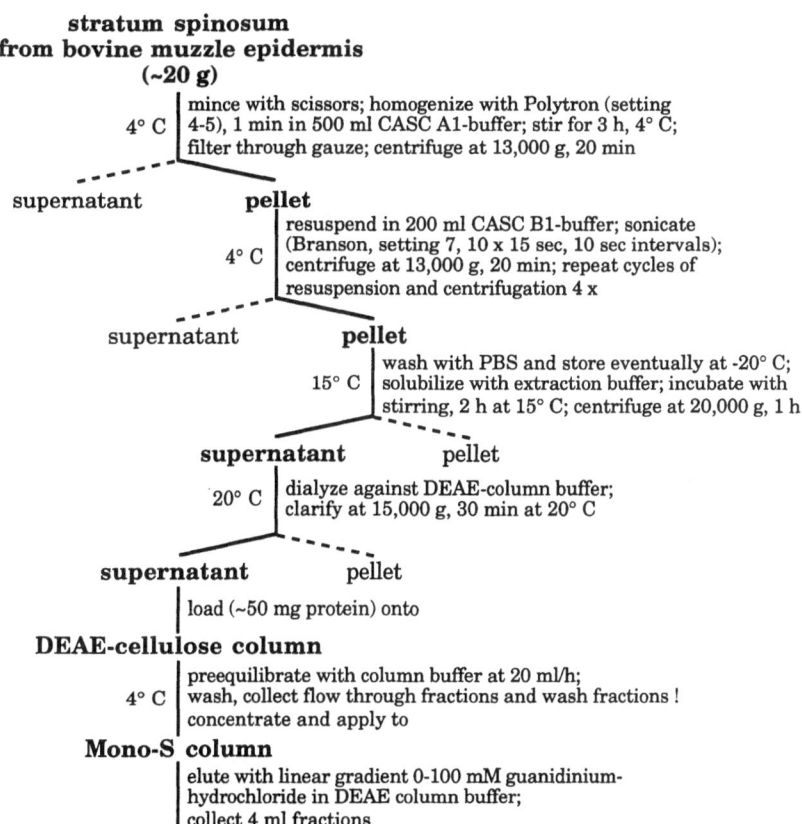

**stratum spinosum
from bovine muzzle epidermis
(~20 g)**

4° C | mince with scissors; homogenize with Polytron (setting 4-5), 1 min in 500 ml CASC A1-buffer; stir for 3 h, 4° C; filter through gauze; centrifuge at 13,000 g, 20 min

supernatant **pellet**

4° C | resuspend in 200 ml CASC B1-buffer; sonicate (Branson, setting 7, 10 x 15 sec, 10 sec intervals); centrifuge at 13,000 g, 20 min; repeat cycles of resuspension and centrifugation 4 x

supernatant **pellet**

15° C | wash with PBS and store eventually at -20° C; solubilize with extraction buffer; incubate with stirring, 2 h at 15° C; centrifuge at 20,000 g, 1 h

supernatant pellet

20° C | dialyze against DEAE-column buffer; clarify at 15,000 g, 30 min at 20° C

supernatant pellet

| load (~50 mg protein) onto

DEAE-cellulose column

4° C | preequilibrate with column buffer at 20 ml/h; wash, collect flow through fractions and wash fractions ! concentrate and apply to

Mono-S column

| elute with linear gradient 0-100 mM guanidinium-hydrochloride in DEAE column buffer; collect 4 ml fractions

BAND 6 POLYPEPTIDE

CINGULIN

A dimeric (2 x 108 kDa) phosphoprotein present in vertebrate tight junctions. Elongated rod-like molecules of 130 x 2 nm.

Source: chicken brush border

Equipment:
- centrifuge
- glass/teflon homogenizer
- sepharose CL-4B (Pharmacia) column (95 x 5 cm)

Chemicals: Na/PO$_4$, EDTA, sucrose, NaCl, NaN$_3$, EGTA, Tris-HCl, Mg-ATP, MgCl$_2$, imidazole, DTT, citric acid, (NH$_4$)$_2$SO$_4$

protease inhibitors: PMSF (0.2 mM), aprotinin (0.1%) and 1 ml/l of 1% chick ovalbumin trypsin inhibitor, TAME, pepstatin A and BAEE (N-α-benzoyl-L-arginine methyl ester)

Have ready: **solution I:** 0.1 M sucrose, 140 mM NaCl, 10 mM sodium phosphate, 10 mM EDTA, 1 mM NaN$_3$, pH 7.0 + protease inhibitors

EDTA-washing buffer: 5 mM EDTA, 20 mM imidazole, 0.1 mM PMSF, pH 7.3

solution II: 50 mM NaCl, 0.1 mM MgCl$_2$, 1 mM EGTA, 10 mM imidazole, pH 7.3

buffer I: 20 mM NaCl, 25 mM imidazole, 10 mM MgCl$_2$, 1 mM EGTA, 0.1 mM DTT, pH 6.5 + protease inhibitors

extraction buffer: 0.6 M NaCl, 0.3 M sucrose, 5 mM EGTA, 25 mM Tris-HCl, 10 mM sodium phosphate, 0.1 mM DTT, pH 7.5 + protease inhibitors

buffer C: 100 mM NaCl, 5 mM sodium phosphate, 10 mM Tris-HCl, 3 mM DTT, 1 mM MgCl$_2$, 1 mM NaN$_3$, pH 7.5

1 M citric acid, cold

1 M Tris-HCl, cold

sepharose 4B column buffer: 0.9 M NaCl, 2 mM EDTA, 0.1 mM EGTA, 0.2 mM DTT, 0.5 mM NaN$_3$, 25 mM Tris-HCl, pH 7.5

Reference: Citi, S. et al. (1989) J. Cell Sci. 93, 107-122

chicken brush border cells (~200 g wet weight)

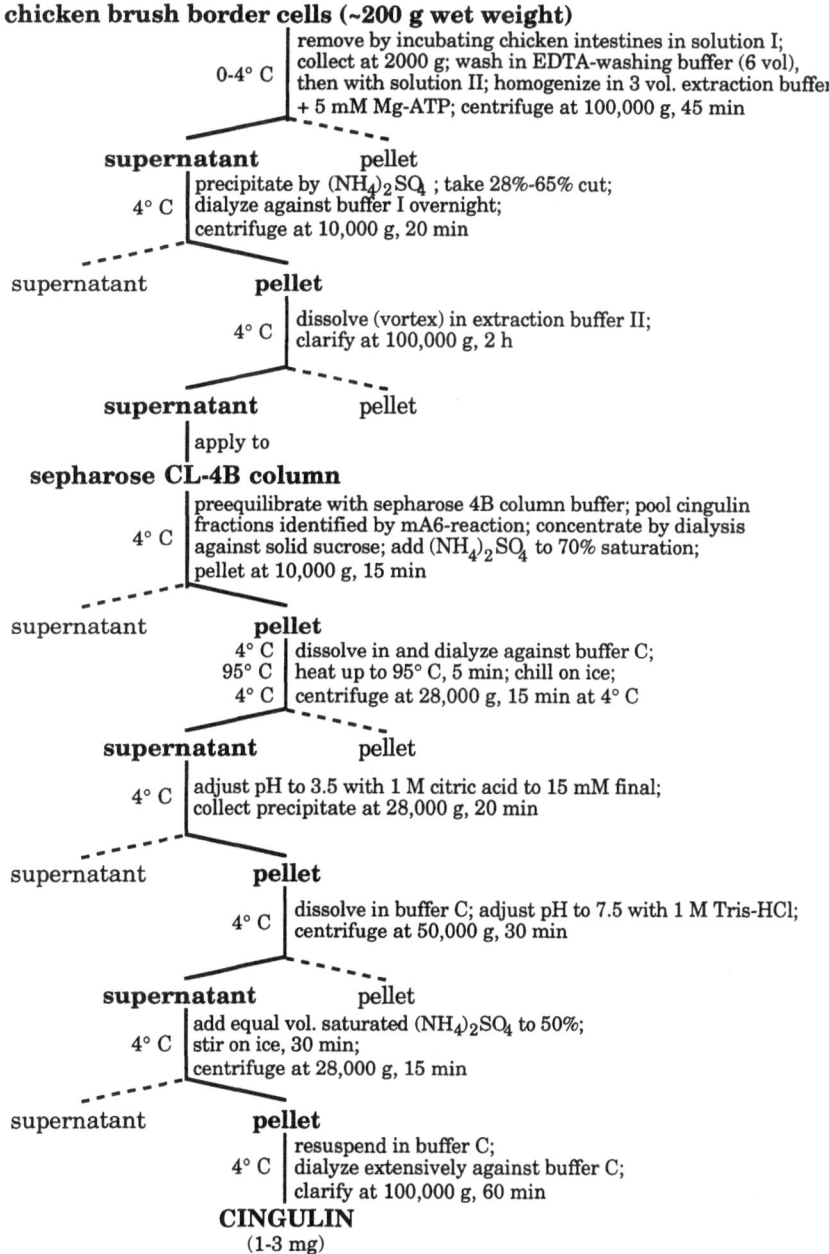

| 0-4° C | remove by incubating chicken intestines in solution I; collect at 2000 g; wash in EDTA-washing buffer (6 vol), then with solution II; homogenize in 3 vol. extraction buffer + 5 mM Mg-ATP; centrifuge at 100,000 g, 45 min |

supernatant   pellet

| 4° C | precipitate by $(NH_4)_2SO_4$; take 28%-65% cut; dialyze against buffer I overnight; centrifuge at 10,000 g, 20 min |

supernatant   **pellet**

| 4° C | dissolve (vortex) in extraction buffer II; clarify at 100,000 g, 2 h |

supernatant   pellet

apply to

sepharose CL-4B column

| 4° C | preequilibrate with sepharose 4B column buffer; pool cingulin fractions identified by mA6-reaction; concentrate by dialysis against solid sucrose; add $(NH_4)_2SO_4$ to 70% saturation; pellet at 10,000 g, 15 min |

supernatant   **pellet**

4° C	dissolve in and dialyze against buffer C;
95° C	heat up to 95° C, 5 min; chill on ice;
4° C	centrifuge at 28,000 g, 15 min at 4° C

supernatant   pellet

| 4° C | adjust pH to 3.5 with 1 M citric acid to 15 mM final; collect precipitate at 28,000 g, 20 min |

supernatant   **pellet**

| 4° C | dissolve in buffer C; adjust pH to 7.5 with 1 M Tris-HCl; centrifuge at 50,000 g, 30 min |

supernatant   pellet

| 4° C | add equal vol. saturated $(NH_4)_2SO_4$ to 50%; stir on ice, 30 min; centrifuge at 28,000 g, 15 min |

supernatant   **pellet**

| 4° C | resuspend in buffer C; dialyze extensively against buffer C; clarify at 100,000 g, 60 min |

CINGULIN

(1-3 mg)

CLATHRIN

The major coat-protein of coated vesicles, trimeric (triskelion) structure consisting of three heavy chains of 180 kDa and three light chains of ~35 kDa.

Source:	bullock brain
Equipment:	• centrifuge, MSE 10x10 rotor
	• Waring blender
	• sucrose gradient maker
	• sepharose CL-4B (Pharmacia) column (45 x 100 cm)
Chemicals:	HEPES-NaOH, Tris-HCl, sucrose, Triton X-100, NaCl, EGTA, $MgCl_2$, NaN_3, EDTA, 2-mercaptoethanol, $(NH_4)_2SO_4$
	protease inhibitors: PMSF
Have ready:	**buffer D:** 10 mM HEPES-NaOH, pH 7.2, 0.15 M NaCl, 1 mM EGTA, 0.5 mM $MgCl_2$, 0.02% NaN_3, 0.2 mM PMSF
	buffer B: 1 M Tris-HCl, pH 7.0, 1 mM EDTA, 0.1% 2-mercaptoethanol, 0.02% NaN_3, 0.2 mM PMSF
	cold saturated $(NH_4)_2SO_4$
Reference:	Pearse, B.M.F. and Robinson, M.S. (1984) EMBO J. 3, 1951-1957
	Gen-Bank:
	J03583 (rat clathrin heavy chain)
	Y00265 (rat clathrin light chains)
	X52272 (yeast clathrin light chains)

bullock brains

$\begin{array}{l}\text{0-4}°\text{ C}\end{array}$ | homogenize in equal vol. buffer D with
Waring blender (3 x 10 s, full speed)
centrifuge at 20,000 g, 30 min

supernatant pellet

RT | clarify at RT;
add 20% Triton X-100 to 1% final conc.;
centrifuge at 100,000 g, 1 h

supernatant **pellet**

4° C | resuspend in minimum buffer D; apply to 50 ml
5-25% sucrose gradient in buffer D + 1% Triton X-100;
centrifuge at 45,000 g, 1 h

supernatant pellet

0-4° C | discard top 10 ml !; save 40 ml supernatant (each);
dilute 3-fold with buffer B;
centrifuge at 100,000 g, 1 h

supernatant **pellet**

4° C | resuspend in small vol. buffer D;
add equal vol. of 2 x buffer B;
centrifuge at 50,000 rpm, 1.5 h in MSE 10x10 rotor

supernatant pellet

load onto

sepharose CL-4B column

4° C | preequilibrate and run in 0.5 M Tris-HCl, 1 mM EDTA,
0.02% NaN_3, 0.2 mM PMSF, pH 7.0; pool clathrin
fractions; concentrate by adding 1 vol. saturated
$(NH_4)_2SO_4$; precipitate at 40,000 g, 30 min

supernatant **pellet**

redissolve in small vol. buffer B;
dialyze against 2 x 100 vol. buffer B;
centrifuge at 50,000 rpm, 1 h in MSE 10x10 rotor

CLATHRIN

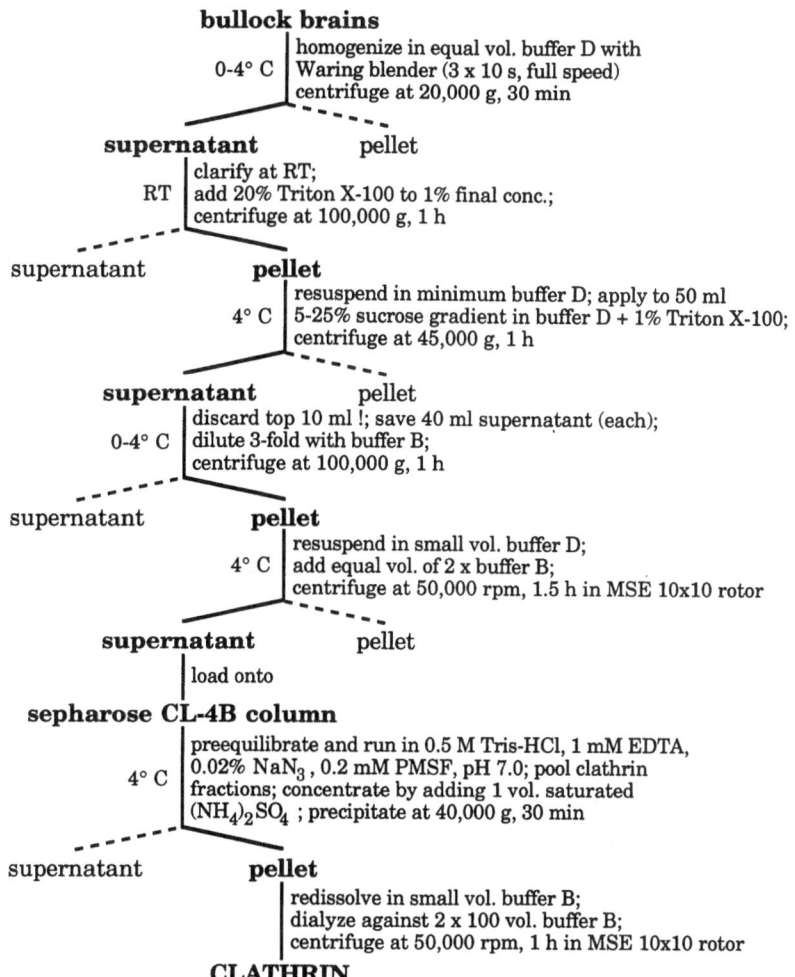

CLATHRIN ADAPTOR PROTEINS

HA-2 is a heterotetramer consisting of α/β-adaptins (110-112 kDa) and α/β-subunits (50/17 kDa), associated with endocytic clathrin coated membranes.
HA-1 is a heterodimer consisting of β'/γ-adaptins (115/104 kDa) and β'/γ subunits (47/19 kDa), associated with Gilgi membranes.

Source: calve brain

Equipment:
- centrifuge, GSA-, SS-34, Sorvall A641, T865.1 rotors
- Waring blender
- glass/glass Dounce homogenizer
- sephacryl S-1000 gel filtration column (2.6 x 60 cm)
- sepharose CL-4B column (2.6 x 90 cm)
- HA-ultrogel (Pharmacia/LKB) column (1.0 x 6.7 cm)

Chemicals: MES, EGTA, $MgCl_2$, DTT, sucrose, Ficoll, Tris-HCl, NaCl, K_2HPO_4, glycerol
protease inhibitors: PMSF

Have ready: **isolation buffer:** 0.1 M MES, pH 6.5, 1 mM EGTA, 0.5 mM $MgCl_2$, 0.2 mM DTT, 0.005% PMSF

Tris-buffer: 1.0 M Tris-HCl, pH 7.0, 0.2 mM DTT

HA-column buffer: 10 mM K_2HPO_4, pH 8.4, 0.1 M NaCl, 10% glycerol, 0.2 mM DTT

Reference: Manfredi, J.J. and Bazari, W.L. (1987) J. Biol.
Chem.   262, 12182-12188
Gen-Bank:
X14971, X14972
       (HA-2 α-adaptin, $α_c$ adaptor subunits)
J04527 (rat β type subunit)
M34175 (β-subunit human fibroblast)
M34176 (β-subunit rat)
M23674 (50 kDa subunit HA-2)
X54424 (HA-2 γ, rat)

246

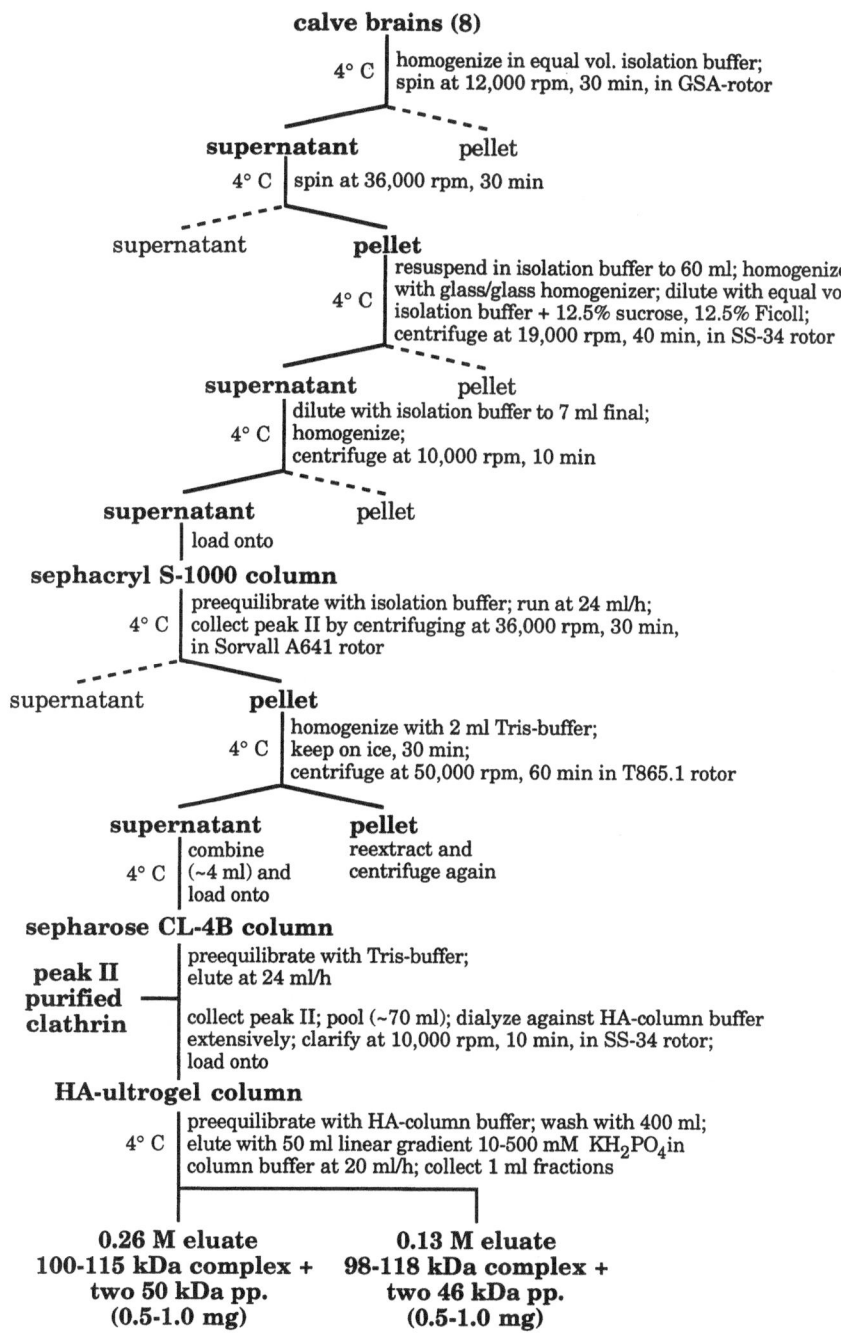

calve brains (8)

4° C — homogenize in equal vol. isolation buffer;
spin at 12,000 rpm, 30 min, in GSA-rotor

supernatant — pellet

4° C — spin at 36,000 rpm, 30 min

supernatant — **pellet**

4° C — resuspend in isolation buffer to 60 ml; homogenize with glass/glass homogenizer; dilute with equal vol. isolation buffer + 12.5% sucrose, 12.5% Ficoll; centrifuge at 19,000 rpm, 40 min, in SS-34 rotor

supernatant — pellet

4° C — dilute with isolation buffer to 7 ml final; homogenize; centrifuge at 10,000 rpm, 10 min

supernatant — pellet

load onto

sephacryl S-1000 column

4° C — preequilibrate with isolation buffer; run at 24 ml/h; collect peak II by centrifuging at 36,000 rpm, 30 min, in Sorvall A641 rotor

supernatant — **pellet**

4° C — homogenize with 2 ml Tris-buffer; keep on ice, 30 min; centrifuge at 50,000 rpm, 60 min in T865.1 rotor

supernatant — **pellet**

4° C — combine (~4 ml) and load onto / reextract and centrifuge again

sepharose CL-4B column

peak II purified clathrin — preequilibrate with Tris-buffer; elute at 24 ml/h

collect peak II; pool (~70 ml); dialyze against HA-column buffer extensively; clarify at 10,000 rpm, 10 min, in SS-34 rotor; load onto

HA-ultrogel column

4° C — preequilibrate with HA-column buffer; wash with 400 ml; elute with 50 ml linear gradient 10-500 mM KH_2PO_4 in column buffer at 20 ml/h; collect 1 ml fractions

0.26 M eluate
100-115 kDa complex +
two 50 kDa pp.
(0.5-1.0 mg)

0.13 M eluate
98-118 kDa complex +
two 46 kDa pp.
(0.5-1.0 mg)

β-COP

β-COP, a 110 kDa protein associated with α (160 kDa), γ (98 kDa), δ (61 kDa) and 35/36 kDa as well as 20 kDa protein as "non clathrin coated vesicle protein complex" in Golgi membranes.

Source:	bovine brain
Equipment:	• centrifuge
	• Waring blender
	• Dounce homogenizer
	• DEAE-cellulose column (700 ml)
	• hydroxylapatide column (140 ml)
	• Mono-Q (Pharmacia) anion exchange column (8 ml)
	• superose-6 (pharmacia) gel filtration column (1.6 x 60 cm)
Chemicals:	Tris-HCl, DTT, KCl, sucrose, EGTA, glycerol, KH_2PO_4, MES, EDTA, $(NH_4)_2SO_4$
	protease inhibitors: PMSF, phenanthroline, leupeptin, pepstatin A, aprotinin
Have ready:	**homogenization buffer:** 25 mM Tris-HCl, pH 8.0, 500 mM KCl, 250 mM sucrose, 2 mM EGTA, 1 mM DTT, 1 mM PMSF, 0.5 mM 1,10-phenanthroline, 2 μM pepstatin A, 2 μg/ml aprotinin, 0.5 μg/ml leupeptin
	dialysis buffer I: 25 mM Tris-HCl, pH 7.4, 50 mM KCl, 1 mM DTT
	resuspension buffer: 25 mM Tris-HCl, pH 7.4, 1 mM DTT, 10% glycerol
	500 mM KH_2PO_4
	hydroxylapatide column buffer: 200 mM KCl, 25 mM Tris-HCl, 1 mM DTT, 10 mM KH_2PO_4, 10% glycerol, pH 7.4
	dialysis buffer II: 25 mM MES, pH 5.8, 1 mM DTT, 10% glycerol
	resuspension buffer: 25 mM Tris-HCl, pH 8.0, 1 M KCl, 1 mM DTT, 10% glycerol
Reference:	Waters, M.G. et al. (1991) Nature 349, 248-251
	Gen-Bank: X57228 (rat β-COP)

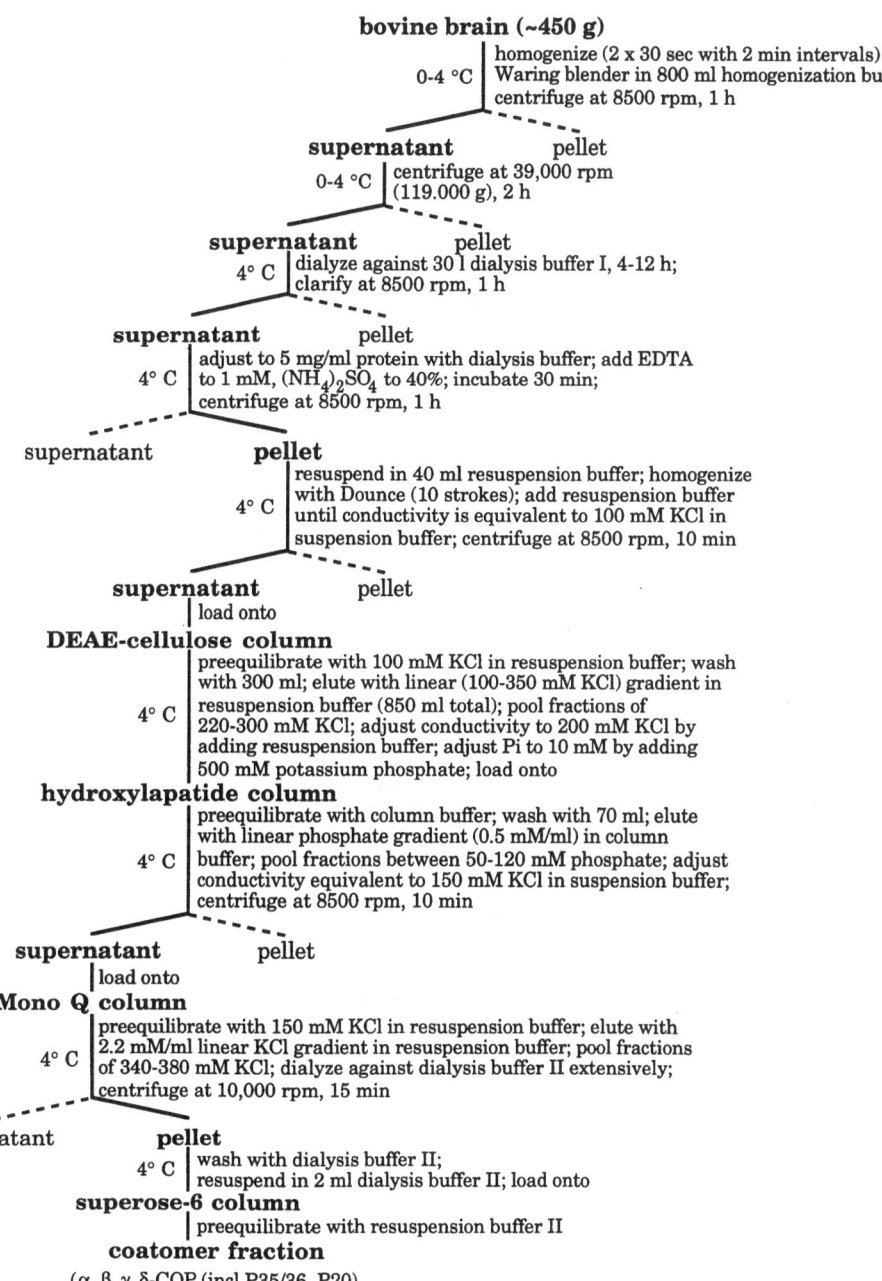

bovine brain (~450 g)

0-4 °C | homogenize (2 x 30 sec with 2 min intervals) with Waring blender in 800 ml homogenization buffer; centrifuge at 8500 rpm, 1 h

supernatant   pellet

0-4 °C | centrifuge at 39,000 rpm (119.000 g), 2 h

supernatant   pellet

4° C | dialyze against 30 l dialysis buffer I, 4-12 h; clarify at 8500 rpm, 1 h

supernatant   pellet

4° C | adjust to 5 mg/ml protein with dialysis buffer; add EDTA to 1 mM, $(NH_4)_2SO_4$ to 40%; incubate 30 min; centrifuge at 8500 rpm, 1 h

supernatant   **pellet**

4° C | resuspend in 40 ml resuspension buffer; homogenize with Dounce (10 strokes); add resuspension buffer until conductivity is equivalent to 100 mM KCl in suspension buffer; centrifuge at 8500 rpm, 10 min

supernatant   pellet

| load onto

DEAE-cellulose column

4° C | preequilibrate with 100 mM KCl in resuspension buffer; wash with 300 ml; elute with linear (100-350 mM KCl) gradient in resuspension buffer (850 ml total); pool fractions of 220-300 mM KCl; adjust conductivity to 200 mM KCl by adding resuspension buffer; adjust Pi to 10 mM by adding 500 mM potassium phosphate; load onto

hydroxylapatide column

4° C | preequilibrate with column buffer; wash with 70 ml; elute with linear phosphate gradient (0.5 mM/ml) in column buffer; pool fractions between 50-120 mM phosphate; adjust conductivity equivalent to 150 mM KCl in suspension buffer; centrifuge at 8500 rpm, 10 min

supernatant   pellet

| load onto

Mono Q column

4° C | preequilibrate with 150 mM KCl in resuspension buffer; elute with 2.2 mM/ml linear KCl gradient in resuspension buffer; pool fractions of 340-380 mM KCl; dialyze against dialysis buffer II extensively; centrifuge at 10,000 rpm, 15 min

supernatant   **pellet**

4° C | wash with dialysis buffer II; resuspend in 2 ml dialysis buffer II; load onto

superose-6 column

| preequilibrate with resuspension buffer II

coatomer fraction

(α, β, γ, δ-COP (incl P35/36, P20), ~0.5 mg per 4 g cytosolic extract)

DESMOPLAKIN I and II

> **Desmoplakin I (~150 kDa) and II (~215 kDa), two major desmosomal plaque proteins.**

Source:	pig tongue
Equipment:	• centrifuge, JA-14, Ti-45 rotors • cheese cloth • keratome • Polytron homogenizer • Branson sonifier • DEAE-cellulose (DE-52, Whatman) column (150 ml) • superose 6 (Pharmacia) column (2.5 x 90 cm) • DEAE-cellulose (DE-52, Whatman) column (5 ml)
Chemicals:	citric acid/sodium citrate buffer, Nonidet P-40, DTT, Tris-HCl, Tris-base, DNA-se I, RNA-se A, $MgCl_2$, EDTA, Tween-20, glycine, NaN_3, Na/PO_4, NaBr, aquacite II (Calbiochem), glycerol **protease inhibitors:** PMSF, DFP, pepstatin A, leupeptin
Have ready:	**homogenization buffer:** 0.1 M citric acid/sodium citrate buffer, pH 2.6, 1 mM DTT, 0.05% Nonidet P-40, DFP (1:10,000), 0.2 mM PMSF, 5 µg/ml pepstatin A, 5 µg/ml leupeptin
	Tris-buffer: 0.1 M Tris-HCl, 1 mM EDTA, 1 mM DTT, 0.01% Nonidet P-40, 5 µg/ml leupeptin, 5 µg/ml pepstatin A
	ice-cold 1 M Tris-base
	Tris-HCl buffer 0.1 M, pH 8.0
	0.1 M EDTA, pH 8.0
	EDTA-buffer: 0.1 mM EDTA, 0.1 mM DTT, 0.01% Tween-20, pH 8.0
	extraction buffer: 1 mM EDTA, 10 mM glycine, 1 mM DTT, 0.05% Tween-20, 1 mM NaN_3, 10 mM Na/PO_4, pH 8.0
	DEAE-column buffer: 4 M urea, 1 mM EDTA, 10 mM glycine, 1 mM DTT, 0.05% Tween-20, 1 mM NaN_3, 10 mM Na/PO_4, pH 8.0
	superose 6 column buffer: 1 M NaBr, 1 mM DTT, 1 mM EDTA, 1 mM NaN_3, 0.05% Tween-20, 10 mM Na/PO_4, pH 7.4
	dialyzing buffer: 10 mM Na/PO_4, pH 8.0, 10% glycerol, 1 mM EDTA, 1 mM NaN_3, 1 mM DTT

Reference: O´Keefe, E.J.O. et al. (1989) J. Biol. Chem. 264, 8310-8318

Gen-Bank: M77830

pig tongues (10-12)

0-4° C | remove epithelium with keratome; mince with scissors; stir in 1.2 l homogenization buffer; homogenize with Polytron, brief bursts, half speed; stir at 4° C, 3 h; filter through cheese cloth; centrifuge at 14,000 rpm in JA-14 rotor, 20 min

supernatant ··· **pellet**

4° C | resuspend in 300 ml citric acid/sodium citrate buffer + 0.01% Nonidet P-40 with syringe; sonify with Branson, full power, 5 x 20 sec with intervals; centrifuge at 1400 g, 20 min

supernatant ··· pellet

| aspirate and centrifuge at 14,000 rpm, 15 min

supernatant ··· **pellet**

| resuspend in same buffer (discard translucent lower part pellet); sonicate; centrifuge and wash 2 x, finally resuspend in 80 ml Tris-buffer, titrate to pH 8.0 with Tris-base; add 0.2 mM PMSF and DFP (1:10,000); incubate on ice, 15 min; add up to 200 ml with 0.1 M Tris-HCl buffer, pH 8.0; centrifuge at 14,000 rpm, 20 min

pellet ··· supernatant

| resuspend with sonication in Tris-HCl buffer + $MgCl_2$ (2 mM final conc); incubate on ice, 2 h; add 1.2 ml 0.1 M EDTA pH 8.0; make up 200 ml with EDTA-buffer; centrifuge at 14,000 rpm, 15 min

supernatant ··· **pellet**

| resuspend in EDTA-buffer; centrifuge at 14,000 rpm, 15 min

pellet ··· supernatant

| resuspend in EDTA-buffer; incubate at 37° C, 30 min; centrifuge at 39,000 g, 15 min

supernatant ··· **pellet**

| resuspend with sonication to 45 ml total + extraction buffer; add eq. vol. of 8 M urea in extraction buffer, slowly with cooling below 5° C; incubate on ice, 16-20 h; centrifuge at 37,000 rpm in Ti-45 rotor, 1 h

supernatant ··· pellet

| aspirate, load onto

DEAE-cellulose column

| equilibrate with DEAE-column buffer + 0.5 M NaBr; wash; elute (10 ml/h) with column buffer + 80 mM NaBr; pool desmoplakin fractions; concentrate to 15 ml with aquacite II; load onto

superose 6 column

| preequilibrate with S-6 column buffer; elute at 20 ml/h

desmoplakin I and II fractions

(Use following procedure to further purify each fraction separately.)

| pool, dialyze against 2 x 1.5 l dialyzing buffer, 20 h; centrifuge at 30,000 rpm, 30 min, in Ti-45 rotor; load onto

DEAE-cellulose column

| preequilibrate with dialyzing buffer (15 ml/h); wash; elute at 2 ml/h with dialyzing buffer + 0.2 M NaBr; collect 1.5 ml fractions; pool; divide into 0.5 ml samples; freeze and store at -80° C

DESMOPLAKIN I or II

58 K

| **A 58 kDa microtubular binding protein on Golgi membranes.** |

Source: rat liver

Equipment:
- Polytron homogenizer
- centrifuge, Ti-45, Ti-60 rotors (Beckman Instr.)
- 0.45 μm filters
- 4000 SW spherogel TSK (Beckman) gel filtration HPLC column (7.8 x 300 cm)

Chemicals: PIPES, EGTA, $MgSO_4$, DTT, sucrose, Triton X-100, KCl

protease inhibitors: 10 μg/ml each leupeptin, pepstatin A, 1 mM PMSF

Have ready: **PEM-buffer:** 0.1 M PIPES, pH 6.8, 1 mM $MgSO_4$, 1 mM EGTA, 1 mM DTT + protease inhibitors

taxol assembled microtubules (see purification scheme p.156/157)

Reference: Bloom, G.S. and Brashear, T.A. (1989) J. Biol. Chem. 264, 16083-16092

7 rat livers (~100 g)

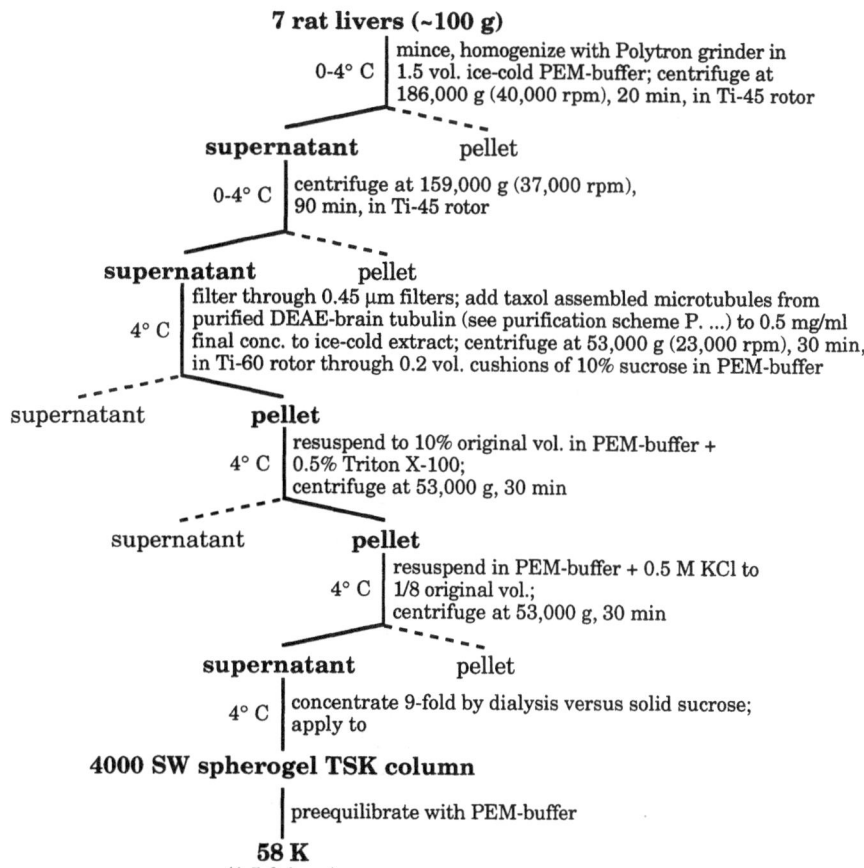

0-4° C | mince, homogenize with Polytron grinder in 1.5 vol. ice-cold PEM-buffer; centrifuge at 186,000 g (40,000 rpm), 20 min, in Ti-45 rotor

supernatant   pellet

0-4° C | centrifuge at 159,000 g (37,000 rpm), 90 min, in Ti-45 rotor

supernatant   pellet

4° C | filter through 0.45 μm filters; add taxol assembled microtubules from purified DEAE-brain tubulin (see purification scheme P. ...) to 0.5 mg/ml final conc. to ice-cold extract; centrifuge at 53,000 g (23,000 rpm), 30 min, in Ti-60 rotor through 0.2 vol. cushions of 10% sucrose in PEM-buffer

supernatant   **pellet**

4° C | resuspend to 10% original vol. in PEM-buffer + 0.5% Triton X-100; centrifuge at 53,000 g, 30 min

supernatant   **pellet**

4° C | resuspend in PEM-buffer + 0.5 M KCl to 1/8 original vol.; centrifuge at 53,000 g, 30 min

supernatant   pellet

4° C | concentrate 9-fold by dialysis versus solid sucrose; apply to

4000 SW spherogel TSK column

| preequilibrate with PEM-buffer

58 K
(1.5-2.0 mg)

TENUIN

A non-actin binding 400 kDa high molecular weight protein localized in adherens junctions and microfilament bundles, rod-like molecules of 400 nm in length.

Source: rat liver adherens junctions

Equipment:
- centrifuge
- Dounce homogenizer
- DEAE-cellulose column (1 x 5 cm)
- sepharose CL-4B (Pharmacia) column (2 x 45 cm)

Chemicals: NaHCO$_3$, sucrose, KCl, MgCl$_2$, HEPES, NP-40, Tris-HCl, EGTA, DTT
protease inhibitors: leupeptin, PMSF

Have ready: **hypotonic solution:** 1 mM NaHCO$_3$, 2 µg/ml leupeptin, pH 7.5

NP-40 solution: 100 mM KCl, 1 mM MgCl$_2$, 10 mM HEPES, 0.1% NP-40, pH 7.5

extraction solution: 2 mM Tris-HCl, 1 mM EGTA, 0.5 mM PMSF, pH 9.2

buffer A: 10 mM HEPES, 1 mM EGTA, 0.1 mM DTT, 1 µg/ml leupeptin, 1 mM PMSF, pH 7.5

Reference: Tsukita, S. et al. (1989) J. Cell Biol. 109, 2905-2915

254

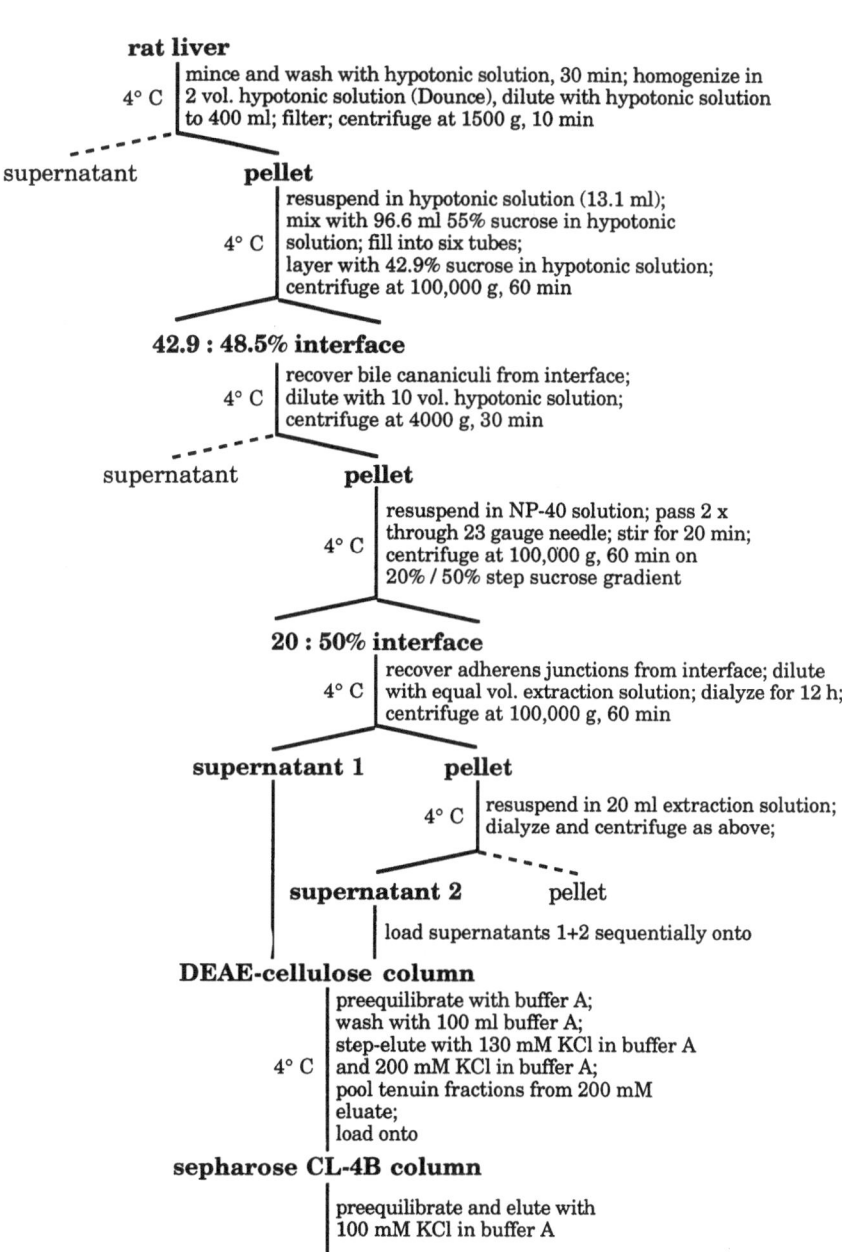

rat liver

4° C | mince and wash with hypotonic solution, 30 min; homogenize in 2 vol. hypotonic solution (Dounce), dilute with hypotonic solution to 400 ml; filter; centrifuge at 1500 g, 10 min

supernatant   **pellet**

4° C | resuspend in hypotonic solution (13.1 ml); mix with 96.6 ml 55% sucrose in hypotonic solution; fill into six tubes; layer with 42.9% sucrose in hypotonic solution; centrifuge at 100,000 g, 60 min

42.9 : 48.5% interface

4° C | recover bile cananiculi from interface; dilute with 10 vol. hypotonic solution; centrifuge at 4000 g, 30 min

supernatant   **pellet**

4° C | resuspend in NP-40 solution; pass 2 x through 23 gauge needle; stir for 20 min; centrifuge at 100,000 g, 60 min on 20% / 50% step sucrose gradient

20 : 50% interface

4° C | recover adherens junctions from interface; dilute with equal vol. extraction solution; dialyze for 12 h; centrifuge at 100,000 g, 60 min

supernatant 1   **pellet**

4° C | resuspend in 20 ml extraction solution; dialyze and centrifuge as above;

supernatant 2   pellet

load supernatants 1+2 sequentially onto

DEAE-cellulose column

4° C | preequilibrate with buffer A; wash with 100 ml buffer A; step-elute with 130 mM KCl in buffer A and 200 mM KCl in buffer A; pool tenuin fractions from 200 mM eluate; load onto

sepharose CL-4B column

preequilibrate and elute with 100 mM KCl in buffer A

TENUIN

ZO-1

An asymmetric, monomeric phosphoprotein of 235 kDa associated with tight junctions and zonulae occludentes in various epithelial cells.

Source: mouse liver

Equipment:
- centrifuges, two SW-28 rotors, SW-41 rotor (Beckman Instr.)
- S-400 gel permeation (Pharmacia) column (1.5 x 48 cm)
- CNBr-activated sepharose /mAb affinity column

Chemicals: $NaHCO_3$, DTT, EDTA, Tris-HCl, sucrose, imidazole, urea
protease inhibitors: leupeptin, chymostatin, pepstatin, soybean trypsin inhibitor, aprotinin, PMSF

Have ready: **homogenization buffer:** 1 mM $NaHCO_3$, pH 8.0, 2.5 µg/ml leupeptin, 0.5 µg/ml chymostatin, 0.5 µg/ml pepstatin, 30 µg/ml soybean trypsin inhibitor, 0.1 trypsin inhibitor units (TiU)/ml aprotinin, 0.2 mM DTT

buffer A: 10 mM Tris-HCl, pH 8.0, 1 mM EDTA, 1 mM DTT + 2.5 µg/ml leupeptin, 0.5 µg/ml chymostatin + pepstatin, 0.02 TiU aprotinin

10:50% sucrose gradient in: 10 mM imidazole, pH 7.4, 1 mM EDTA, 0.2 mM PMSF

Reference: Anderson, J.M. et al. (1988) J. Cell Biol. 106, 1141-1149

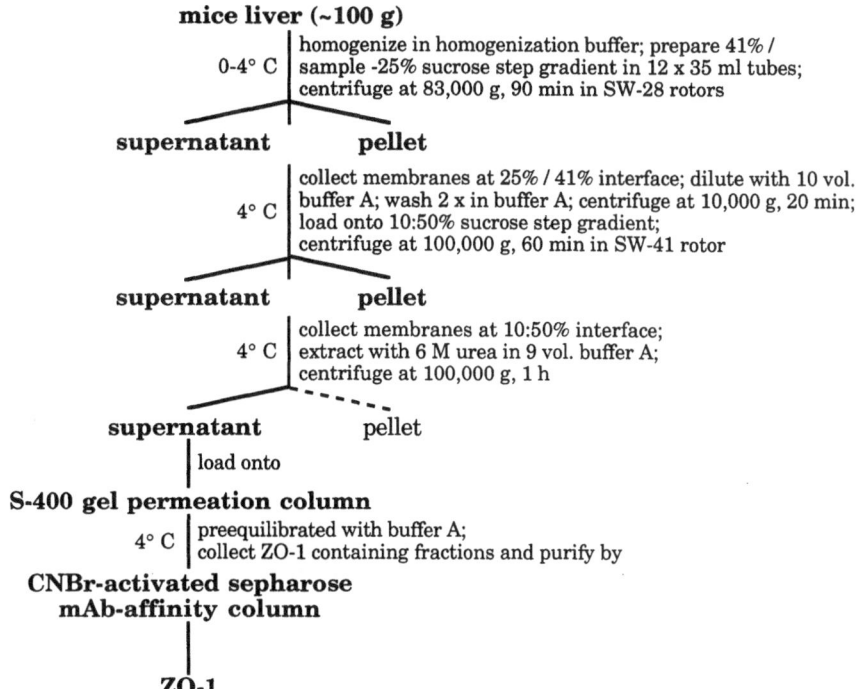

mice liver (~100 g)

0-4° C | homogenize in homogenization buffer; prepare 41% /
sample -25% sucrose step gradient in 12 x 35 ml tubes;
centrifuge at 83,000 g, 90 min in SW-28 rotors

supernatant   **pellet**

4° C | collect membranes at 25% / 41% interface; dilute with 10 vol.
buffer A; wash 2 x in buffer A; centrifuge at 10,000 g, 20 min;
load onto 10:50% sucrose step gradient;
centrifuge at 100,000 g, 60 min in SW-41 rotor

supernatant   **pellet**

4° C | collect membranes at 10:50% interface;
extract with 6 M urea in 9 vol. buffer A;
centrifuge at 100,000 g, 1 h

supernatant   pellet

load onto

S-400 gel permeation column

4° C | preequilibrated with buffer A;
collect ZO-1 containing fractions and purify by

**CNBr-activated sepharose
mAb-affinity column**

ZO-1

C. Suppliers List

Beckman Instruments

Addresses

Sales and Service Offices:

Beckman Instruments, Inc.
2500 Harbor Boulevard
Fullerton, CA 92634
U.S.A.

Order Assistance: United States 800-742-2345

Field Service: United States 800-551-1150
Alaska (Call Collect) (415) 857-0750
Hawaii (800) 396-4233

Worldwide Sales Offices:

Germany
Beckman Instruments GmbH
Frankfurter Ring 115
80807 München
Federal Republic of Germany
Telephone: (49) 89-3887-1

Australia
Beckman Instruments (Australia) Pty.Ltd.
24 College Street
Gladesville, NSW 2111
Australia
Telephone: (61) 02 816-5288

Canada
Beckman Instruments (Canada) Inc.
1045 Tristar Drive
Mississauga, Ontario
Canada L5T 1W5
Telephone: (416) 670-1234

France
Beckman Instruments France S.A.
92/94 Chemin des Bourdons
93220 Gagny, France
Telephone: (33) 1 43 01 70 00

Italy
Beckman Analytical S.p.A.
Centro Direzionale Lombardo
Via Roma 108, Palazzo F/1
20060 Cassina De Pecchi
Milano, Italy
Telephone: (39) 2-953921

Japan
Beckman Instruments (Japan) Ltd.
6, Sanbancho
Chiyoda-Ku, Tokyo 102
Japan
Telephone: (81) 3-3221-5831

Netherlands
Beckman Instruments Nederland B.V.
Nijverheidsweg 21
P.O.Box 47
3640 AA-Mijdrecht
The Netherlands
Telephone: (31) 2979-85651

Spain
Beckman Instruments Espana S.A.
Avda del Llano Castellano 15
Madrid - 28034
Spain
Telephone: (34) 1-358-00-51

Sweden
Beckman Instruments A.B.
Archimedesvaegen 2
Box 65
S-16126 Bromma
Sweden
Telephone: (46) 8-98 5320

Switzerland
Beckman Instruments International S.A.
22, Rue Juste Olivier
1260 Nyon
Switzerland
Telephone: (41) 22-9940 707

United Kingdom
Beckman Instruments (UK) Ltd.
Oakley Court
Kingsmead Business Park
London Road
High Wycombe, Bucks. HP11 1JU
England
Telephone: (44) 1494-441 181

Products

BECKMAN Centrifuges

OPTIMA XL-A	Analytical ultracentrifuges
OPTIMA XL series	Preparative ultracentrifuges
OPTIMA L series	Preparative ultracentrifuges
OPTIMA XL/TLX series	Table Top ultracentrifuges
Airfuge	Table Top ultracentrifuges
Avanti J series	High Performance centrifuges
J2 series	High Speed centrifuges
Avanti 30	Table Top High Performance centrifuges
J6 series	High Capacity centrifuges
GS-6 series	Table Top centrifuges
GS-15 series centrifuges	Laboratory + Microfuge
Microfuges	Micro Table Top centrifuges

BIO-RAD

Adresses

USA
Bio-Rad Laboratories
2000 Alfred Nobel Drive
Hercules, CA 94547
Telephone: (510) 741-1000
            (800) 424-6723
Telefax: (510) 741-1060
            (800) 879-2289

Germany
Bio-Rad Laboratories GmbH
Heidemannstr. 164
D-80939 Munich
Telephone: (089) 318840
Telefax: (089) 31884100

Japan
Nippon Bio-Rad Laboratories
KK.
Sumitomo Seimei Kachidoki
Bldg.
3-6 Kachidoki 5-Chome
Chuo-Ku, Tokyo 104
Telephone: 03-3534-7665
Telefax: 03-3534-8497

Switzerland
Bio-Rad Laboratories AG
Kanalstraße 17
CH-8152 Glattbrugg
Telephone: 01-809 55 55
Telefax: 01-809 55 00

United Kingdom
Bio-Rad Laboratories Ltd.
Maylands Avenue
Hemel Hempstead
Hertfordshire HP2 7TD
Telephone: 01442-232552
Telefax: 01442-259118

Products

ion exchange supports:	Macro-Prep®: Q, S, CM, DEAE 50 µm particle size, 10 µm particle size prepacked in Bio-Scale column (high resolution)
hydrophobic chromatography:	Macro-Prep®: HIC- Support Methyl, t-Butyl, 50 µm Ø

HPLC-grade ammonium sulfate with low UV-absorbance for gradient elution

hydroxylapatide:	Bio-Gel® HTP (crystalline) Macro-Prep® CHT (ceramic, spherical)
affinity chromatography:	Affi-Gel, (argarose based), 10,15, Hz, Protein A, Polymixin Affi-Prep, (polymer based), 10, Hz, Protein A
gel filtration supports:	Bio-Gel P (polyacrylamide based) Bio-Gel A (argarose based)

protein standards for chromatography and electrophoresis

equipment for analytical (PROTEAN® CELLS) and preparative (PREP-CELL® and ROTOFOR) protein purification

equipment and columns for low (Econo System), medium (BioLOGIC) and high pressure chromatography

Pharmacia

Adresses

North America
Pharmacia Biotech
800 Centennial Ave.
P.O. Box 1327
Piscataway, NJ, 08855-1327
Telephone: (800) BIA-2599
Telefax: (908) 457-8666

Japan
Pharmacia Biotech K.K.
Honda Denki Bldg, 5-37
Kami-Osaki 4-chome
Shinagawa-ku
Tokyo 141
Telephone: 3-34929499
Telefax: 3-34924982

England
Pharmacia Biotech Ltd.
23 Grosvenor Road
St. Albans
Herts, AL1 3AW
Telephone: 727-804075
Telefax: 727-814001

Sweden
Pharmacia Head-Office
Rapsgatan 7
S-751 82 Uppsala
Telephone: 18-165700
Telefax: 18-115890

Germany
Pharmacia Biotech GmbH
Munzinger Str. 9
D-79111 Freiburg
Telephone: 761-4903-0
Telefax: 761-4903-249

Products

high resolution chromato-
graphic media for gel-
filtration and ion-exchange
chromatography:

sepharose 4B, 6B
phenyl sepharose
PD-10 gel filtration columns
NAP-10 gel filtration columns
Mono Q FPLC columns
sephadex gel filtration media
sephacryl gel filtration media

Whatman

Adresses

Whatman Lab Sales Ltd.
St. Leonards Rd.
Maidstone, Kent ME16 0LS
Telephone: 1622-676670
Telefax: 1622-677011

Products

high resolution ion exchange resins: DE-52, DE-53
CM-cellulose
P-11 cellulose

INDEX

Cold, Hard Facts For Process Chromatographers

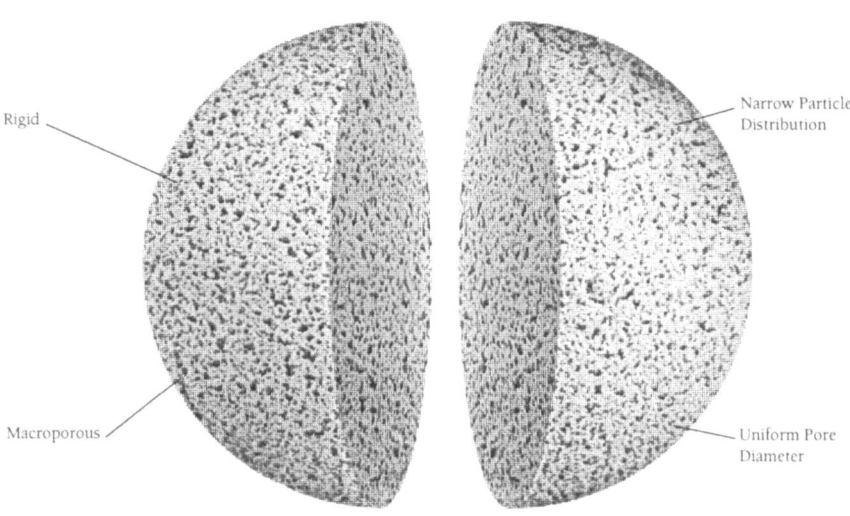

Rigid

Narrow Particle Distribution

Macroporous

Uniform Pore Diameter

Introducing Macro-Prep® supports: a small bead with a big effect on the bottom line

The facts about this unique support speak for themselves.

- High throughput— 3,000 cm/hr flow rate
- No shrinking or swelling
- Reusable, autoclavable, 1 M NaOH stable, ethanol sterilizable
- Batch to batch reproducibility
- Easy scale up, more cost effective

The Macro-Prep supports are available in Q, S, DEAE, CM, HIC (t-butyl and methyl) and Ceramic Hydroxyapatite. They are produced under GMP and have Drug Master Files.

These are the cold hard facts. To receive technical data sheets for Macro-Prep supports, contact your local Bio-Rad office

BIO-RAD

Bio-Rad Laboratories

Life Science Group

(US)(800) 4BIORAD • (AU) 02-805-5000 • (AT) 0222-877 89 01 • (BE) 09-385 55 11 • (CA) (416) 624-0713 • (CN) 2563146 • (FR) 01-49 60 68 34 • (DE) 089 31884-0 • (IT) 02-21609 1 • (JP) 03-3534-7515 • (HK) 7893300 • (NL) 08385-40666 • (NZ) 09-443 3099 • **Scandinavia** 46 (0) 8 590-73489 • (ES) (91) 661 70 85 • (CH) 01-810 16 77 • (GB) 0800 181134   SIG 061793